OIL SHOCK

A REPORT OF THE HARVARD ENERGY SECURITY PROGRAM

Written under the auspices of the Energy and Environmental Policy Center, John F. Kennedy School of Government, Harvard University.

OIL SHOCK
Policy Response and Implementation

edited by
ALVIN L. ALM
ROBERT J. WEINER

BALLINGER PUBLISHING COMPANY
Cambridge, Massachusetts
A Subsidiary of Harper & Row, Publishers, Inc.

International Standard Book Number: 0-88410-900-3

Library of Congress Catalog Card Number: 83-22459

Printed in the United States of America

Library of Congress Cataloging in Publication Data

Main entry under title:

Oil shock.

   Papers presented at a conference held in Cambridge, Mass., 15-16 July 1982, and sponsored by the Harvard Energy and Environmental Policy Center.
   Includes index.
   1. Petroleum industry and trade—Government policy—United States—Congresses. 2. Energy policy—United States—Congresses. 3. Inflation (Finance)—United States—Effect of energy costs on—Congresses. 4. Petroleum—United States—Reserves—Congresses. I. Alm, Alvin L. II. Weiner, Robert J. III. John F. Kennedy School of Government. Energy and Environmental Policy Center.
HD9566.037 1983   333.8'232'0973   83-22459
ISBN 0-88410-900-3

CONTENTS

LIST OF FIGURES

LIST OF TABLES

PREFACE

This book is the product of a conference entitled "Energy Security Policy Implementation Issues," which was held in Cambridge, Massachusetts, July 15–16, 1982. The conference was sponsored by the Harvard Energy and Environmental Policy Center, under the auspices of the Energy Security Program and brought together representatives of the executive and legislative branches of the federal government, business, academia, international organizations, and the press. The purpose of the conference was to present and debate work on energy security policy and implementation being undertaken by the program and elsewhere.

The Harvard Energy Security Program was organized in 1979 jointly by the Energy and Environmental Policy Center and the Center for Science and International Affairs. The broad objective of the program is to support work in energy security studies by gathering researchers from various disciplines. The first effort of the program was *Energy and Security*, edited by David Deese and Joseph Nye, published in 1981.

Since 1981, research has concentrated on proposing and evaluating specific policy responses to the issues raised in that book and elsewhere. The conference was held to exchange views on work done both within and without the program. The reactions of those in attendance encouraged the program's staff to disseminate the research more widely; hence, this volume.

Energy security is a broad, interdisciplinary area that we cannot aspire to cover in a comprehensive fashion in this book. Rather, we have taken three important topics associated with supply disruptions—the behavior of the oil market and the oil industry, the effect on the economy, and the role of the Strategic Petroleum Reserve—and examined policy and attendant implementation issues. In editing this book, we have not endeavored to produce a seamless web of analysis. The differences of opinion and vigorous exchanges among the participants that characterized the conference are reflected here.

Our debts in editing this book are many, but there are a few people to whom we are especially grateful. Cynthia Vissers O'Hanley kept the operation afloat, proofread, and edited most of the book. Carol Donahue coordinated the word processing and reprocessing with great skill and unfailing good humor. Their work was sensational.

Special thanks for organizing and running the conference go to Marla Smith and Patricia Slote. Lisa Burdick, Chris Lundblad, and Leslie Sterling provided able typing. Maja Arnestad's editorial comments were particularly well taken. We greatly appreciate the patience and expeditious replies of the authors. The staff of the Energy and Environmental Policy Center and the Center's Corporate Board of Advisors generously provided time and resources.

Two papers presented at the conference have appeared in similar form in technical journals, which have kindly allowed their reprinting. Chapter 5 is reprinted with permission from the *Natural Resources Journal*. Chapter 11 is reprinted with permission from *Energy Economics*. All the chapters but one are based on conference papers; James Plummer was unable to attend, but his piece complements the other work on stockpiling.

We have benefited from conversations with our professional colleagues, particularly those at the Stanford Energy Modeling Forum, the Center for Energy Policy Research at the Massachusetts Institute of Technology, and the International Energy Agency of the Organization for Economic Cooperation and Development.

We acknowledge the strong financial support of the United States Department of Energy. The support and interest of the department's staff has been an important factor in the rapid progress made over the last ten years in understanding energy markets and their relationships with the macroeconomy. We are especially indebted to William Vaughan, Ronald Winkler, Glenn Coplon, Patricia Knight, Jerome Temchin, and Jerry Blankenship.

Finally, we thank the conference participants. John Hill, John Lichtblau, Stuart Eizenstat, George Eads, Knut Mork, John Treat, and Darius Gaskins were kind enough to serve as session chairmen or discussants. Those who attended succeeded in making it unusually productive through their vigorous participation.

1 INTRODUCTION
Managing Oil Shocks

Alvin L. Alm

Few people see the relationship between the unemployment lines in Detroit and Cleveland in 1983 and the 1978 strike of oil workers in Iran. Most Americans have come to believe that runaway government deficits and tight money policies of the Federal Reserve are the only culprits. Because no immediate connection is made between oil price shocks and later economic distress, most Americans do not consider measures to protect against oil interruptions an important ingredient in maintaining economic health. Yet, as we shall see, the connection between the abstract concepts of energy security and our daily lives is immediate and real.

The two oil price shocks of the 1970s dominated the economic fortunes of the major industrialized countries. In 1972, most of these countries were experiencing healthy economic growth and tolerable levels of inflation. By 1974, all of the economies were in a shambles, with double-digit inflation plaguing all but the Federal Republic of Germany and negative growth afflicting the United States, Japan, and the United Kingdom. By 1978, the Western economies had brought inflation down and restored growth. By 1980, however, members of the Organization for Economic Cooperation and Development (OECD) except for Germany and Japan were again facing

This chapter was written while the author was director of the Harvard Energy Security Program.

double-digit inflation and, except for Italy and Japan, growth rates of less than 2 percent. Ironically, the large oil-consuming nations with the lowest dependence on imports—namely, the United States, the United Kingdom, and Canada—experienced the lowest growth in 1980 (Yergin and Hillenbrand 1982: 377–78).

Some of our current economic woes still go back to these oil price shocks. The monetary policy imposed since 1979 has slowed the economy in general and seriously crippled the automobile and housing industries. That policy was an outgrowth of the failure of fiscal policy to cope with the high levels of inflation arising from the 1979–80 oil price shock. The unprecedented sharp drop in the growth of productivity in 1974 and then a second drop in 1979 implicates energy price shocks as a guilty party in reducing the level of U.S. productivity. Clearly, other factors, such as foreign competition, have also affected U.S. and foreign economic growth. But the strong evidence that energy shocks played a major role is almost undeniable.

What can be done to minimize this economic misery in the future? For most security threats, the antidotes are obvious. For military threats, societies increase armaments and armies. To curb the devastation from floods they build dams, and to prevent outbreaks of disease they inoculate the population. But for energy security there are large differences of opinion about what to do and serious political problems surrounding many of the options. Equally important is the fact that most of the implementation steps have not been worked out either as conceptual models or as workable programs. Without having created effective energy security measures in advance, the United States could be saddled with the same flawed mechanisms that failed so miserably in the past.

A number of excellent studies have examined previous supply interruptions, analyzed their economic impact, and suggested the broad outlines of policy (Deese and Nye 1981; Krapels 1980; Plummer 1982a; Verleger 1982; Yergin and Hillenbrand 1982). This study differs from these efforts by focusing on a few specific policy and implementation issues. Chapters on recycling and stockpile management illuminate some of the issues that policymakers and administrators will face in crafting energy security policies and measures. The study concludes that much can be done to strengthen our energy security but that the lack of political consensus and understanding of implementation obstacles present formidable barriers. Until we learn to combine our theoretical understanding of the energy security issue with an understanding of the less enticing implementation issues, progress will be slow.

This chapter has two purposes: to serve as an introduction and backdrop to the particular issues examined in this book and, on a broader level, to synthesize my thoughts based on experience in government and university settings.

CRUDE OIL ACCESS

Historically, the world oil market was almost the private domain of the major oil companies. They entered into concession agreements that allowed them to explore and produce oil in the major Mideast producing states with little interference. Even though this power was eroding by 1973, the major oil companies were still able to allocate supplies during the Arab oil embargo. By 1979, however, they had lost control of distribution for two reasons. First, their control over crude oil supplies had diminished further. Moreover, the oil companies found that there was a large difference between managing the distribution during the Arab oil embargo—in which all the majors had control over crude oil supplies and could even out the overall cutbacks—and the situation they faced during the Iranian Revolution. By focusing the shortages on certain companies, the Iranian disruption caused cascading cutbacks among companies that ultimately led to tremendous pressures on the spot market and the establishment of new institutional arrangements in the world oil market. For example, the loss of Iranian supplies forced British Petroleum (BP) to cut back sales to third parties by 700 thousand barrels of oil per day. Then Exxon, which had a third-party contract with BP for Iranian supplies, was forced to reduce its third-party sales to Japan. At the end of the line, the Japanese had no choice but to go into the spot market (Verleger 1982: 12).

A number of U.S. independent refiners lost supplies, either directly or indirectly, causing them either to go to the spot market or to the U.S. government for crude oil supplies. At first the U.S. government discouraged spot market purchases, but as the shortage progressed, it dropped this policy, both because the policy was not working and because of rising concern over shortages. The government allocated 215 million barrels of crude oil from fifteen major companies to smaller companies under the Buy/Sell Program, a program designed to even out shortages between integrated and nonintegrated refiners. (Chapter 3 provides details on the program.)

During the course of the Iranian episode, the world oil market changed at a rapidly accelerating pace. More oil became tied up in government-to-government or government-to-company arrangements. More oil was sold with special destination requirements or other political limitations. Saudi Arabia was selling access to crude oil not controlled by ARAMCO by requiring investments in its two industrial complexes at Yanbu and Jubail. The Japanese, stung by cancellation of contracts from major U.S. oil companies, scoured the Middle East for new purchaser arrangements. With the exception of creating many more purchasers in the world oil market, the changes later disappeared as rapidly as they arose.

Plummeting oil demand has temporarily reduced Western dependence on oil imports and vulnerability to supply interruptions. Spare production capacity rose from about 3 million barrels per day (mbd) before the Iranian interruption to a current level of over 10 mbd. Indeed, until demand rises and some of the excess capacity declines, nothing less than a major Middle East war could realistically cause supply interruptions. An interruption during the next few years may well have the greatest impact on Exxon, Standard Oil of California, Texaco, and Mobil—the four partners that constitute ARAMCO.

Even though the market has dramatically changed, concern over access to crude oil supplies remains a major issue in the United States. Indeed, both houses of Congress passed the Standby Petroleum Allocation Act (SPAA) in December 1981, which would have reestablished a system for allocating crude oil supplies during an oil supply interruption. Although that legislation was vetoed by President Reagan and Congress failed to override the veto, the issue has not died. Congress feels much more comfortable with an allocation system, particularly because individual members can take credit for meeting the interests of certain politically powerful groups.

Three arguments are generally advanced to justify continued government intervention during oil supply interruptions. First, proponents of allocations argue that independent refiners would be unable to obtain crude oil supplies and hence would go out of business. Those with a more suspicious outlook believe major oil companies would undersell the independents in order to hasten their economic demise. A second concern, which has powerful political appeal, is that certain regions of the country would be inordinately hard hit, unable to bid supplies away from other areas at any price, or would face exorbitant prices. Finally, some argue that without a government allocation program, refiners cut off from supplies will aggressively bid up prices in the world oil market to the economic loss of all consumers.

The possibility that major oil companies would use the occasion of shortages to squeeze out independent marketers through predatory pricing—by taking markets away from them or even driving them out of business—is widely feared. During an interruption, major oil companies could use the occasion to discipline independent marketers and refiners through denial of crude access, restrictive reprocessing agreements, and product sales restrictions on jobbers. While these possibilities all certainly exist, it is unlikely they would be widespread and lasting. Over the long term, it is inconceivable that major oil companies could drive out enough independents to gain permanent control over pricing. Even at the regional level, permanent eradication of a sufficient number of independent refiners and marketers to affect pricing and competition would not be easy. And this predation would take place in a political environment generally hostile to major oil companies and in a legal environment that could lead to antitrust action or public embarrassment.

While predation is not an immediate threat, this does not mean that major oil companies would necessarily establish pricing policies based on the marginal cost of crude oil or that they would sell supplies freely to all bidders willing to pay the market price. First, companies are likely to face severe political pressure to keep prices down. In the Federal Republic of Germany, considered the bastion of a free energy market, Chancellor Schmidt chided oil companies for provocative profit levels in 1979. Based on behavior during the 1979 crisis, U.S. oil companies could reasonably anticipate jawboning or even punitive regulatory action if consumers perceived they were being gouged by high prices. Second, large oil companies have some propensity to allocate supplies within their organizations and to historic customers by administrative fiat rather than on the basis of the highest prices. This behavior could be observed internationally during the 1973–74 crisis and even domestically in 1979 for products not under direct government allocation. A combination of commercial practice and, some would argue, the legal requirements of the Uniform Commercial Code, led to a strong presumption that during another oil crisis the market would be limited in moving products to the highest bidder. These institutional impediments would make market allocation of crude oil supplies more difficult than would be expected under a classical economic model.

When refiners with high crude costs compete directly with integrated firms with lower priced crude, then the high-cost firms face three options: They can cut prices and lose money, curtail operations, or go out of business. When independent refiners with the same crude oil costs sell in isolated markets, however, they can always pass higher costs on to their customers. In the first case, the refiner may be endangered; in the second, consumers must pay more for petroleum products than consumers elsewhere in the country.

The potential for a large number of independent refiner bankruptcies is not great. During supply interruptions the rising tide of higher profits will help most refiners, which are likely to gain large profits from whatever domestic production they have as well as from oil delivered under contract. Even if some independent refiners go out of business, the refining business is still considerably more competitive than most. The purpose of competition, in any case, is not to encourage the maximum number of competitors but to encourage healthy price competition. Government does not necessarily need to accept the burden of ensuring that no refineries go out of business during supply interruptions any more than it needs to protect other businesses from adverse external threats.

The final problem posed by supply interruptions relates directly to consumers. In some cases, only one refinery serves a region. If cut off from supplies, that refinery would need to buy supplies on the spot market or from other refiners with access to crude oil and would inevitably need to charge more for products than refiners with better access to crude oil supplies. Consumers in these regions may feel they are unfairly disadvantaged, although, in fact, they would not be paying any more than if all companies charged at the marginal cost level. The policy question is whether government has an obligation to ensure that consumer prices are relatively consistent among regions during a supply interruption. If the answer is yes, then government intervention could be justified.

If one believes that the government should ensure consistent prices across all regions, then a pinpointed regulatory program could be designed to deal with only the handful of serious problems, allowing the market to function in all other cases. But is this possible? During an interruption a large number of refiners would importune the government for help, usually backed by local political leaders. Exceptions to any general rule of equity would follow from intolerable political pressures, and these exceptions would become precedents for future cases. Before long it could become easier to grant allocations than to deny them, spreading the tentacles of the allocation program wider and wider. At this point the allocation program itself intensifies shortages, leading to the need for even more allocation. Although a carefully tailored allocation program is a theoretical possibility, its practical feasibility depends on a disciplined political process. Based on history such discipline appears highly unlikely, at least in the United States.

ECONOMICS AND EQUITY

Oil price shocks harm industrialized economies in many ways. In the short term they cut effective demand by transferring wealth to producing nations and induce governments to tighten monetary and fiscal policies to control inflation. Over a longer period of time these economies must adjust to an altered mix of goods and replacement of now obsolete capital, causing declines in productivity and economic growth.

The chief economist of OECD has developed what she calls a "rudimentary order-of-magnitude estimate" of the economic growth losses emanating from the 1979 oil supply interruption. She estimates that income in OECD nations was roughly 5 percent smaller in 1980 and 7.5 percent smaller in 1981 because of the oil price shock, amounting to income losses of $350 billion in 1980 and $570 billion in 1981 (Ostry 1981).

The first category of loss arises from the "terms of trade loss," where billions of dollars of wealth are transferred from consuming to producing nations. Oil and oil-related products become more expensive, reducing the real value of the money income of nonoil producers and correspondingly increasing the income of oil producers. The transfer, representing roughly 2 percent of total OECD income in 1980 and 1981, can be reduced as imported oil demand and oil prices fall. If nothing further happened, this would be the extent of income losses, with further growth occurring much as it did before.

But effective world demand falters because some of the major oil producers cannot immediately spend their increased new revenues. This diminished demand could be reduced if countries borrowed all of these new savings and spent them. The sum of this borrowing, however, turns out to be less than the initial increase in oil producers' savings, causing reductions in total world demand for goods and services. The impact of this induced extra saving is estimated at 2.25 percent of OECD GNP in 1980 and 2.75 percent in 1981 (Ostry 1981).

Finally, governments impose tight monetary and fiscal policies to control inflation. For 1980, this impact is estimated at 0.5 percent, rising to 2.75 percent by 1981. The restrictive policies of the U.S. Federal Reserve, for example, decimated the automobile and housing industries, adding to recessionary conditions in associated industries such as steel and lumber. High U.S. interest rates attracted capital from other countries, causing their central bankers to raise interest rates; by this process, some of the economic distress has been exported.

Oil price shocks have some longer term impacts. Labor and capital are substituted for energy, lowering measured productivity. Demand for oil and oil-related products will decline, changing the composition of total product demand. A great deal of capital will become obsolete because of higher energy prices. The recession attendant to an oil shock serves to cut profits and slow capital stock replacement.

Wage rates are important to the process of adjustment. If wage rates fell as oil prices went up, the reduction of effective demand would not be nearly as serious. If inflation from oil price increases were offset by falling wage rates, restrictive monetary and fiscal policies would be less necessary. In Chapter 5, Pindyck and Rotemberg argue for a reduction in payroll taxes, as well as investment tax credits, to cope with this problem.

In addition to the macroeconomic impacts already discussed, oil price shocks also effect a dramatic redistribution of wealth. As already mentioned, a large shift occurs between consuming and producing nations. But in some OECD countries—notably the United States, Canada, and Great Britain—shifts occur within the country. In the United States, for example, a large oil price increase would initially result in larger government revenues from the corporate income tax and the Windfall Profits Tax. The remainder of these gains would be transferred to lease holders and oil companies. Should the next interruption occur after 1985, when new natural gas prices are decontrolled, natural gas producers will also be major beneficiaries of an oil price shock.

These large transfers of wealth will likely cause an outcry for political action. During and after oil price shocks, newspapers will proclaim record oil company profits while average Americans will see prices soaring and their real income declining. Under these conditions politicians will be under great pressure to do something to prevent "profiteering" by oil companies. Based on experience, reimposition of oil price controls could be almost irresistible, unless some mechanism existed to deal with perceived inequities and severe hardships.

Many theoretically attractive suggestions will face strong political opposition, leaving decisionmakers to their political instincts and ideologies in dealing with distributional changes. The choices that confront them are three: doing nothing, reimposing price controls, and recycling oil revenues. Staying with the status quo could conceivably work if an administration stood ready to veto any congressional legislation and to lobby hard against any attempt to override a presidential veto. But its chances of success are limited by the political pressures arising from a supply interruption.

A second option, reimposition of price controls, would be very attractive politically, as it would simultaneously punish the oil companies and keep consumer prices down. Some even argue that price controls reduce inflation and ease macroeconomic shocks. But the cost of reimposing price controls is high. During the interruption, shortages would develop, resulting in gasoline lines and perhaps shortages in other products. Without the pressure of higher prices on demand, U.S. consumption would be greater, eventually leading to higher world oil prices and greater economic losses. Also, once price controls have been instituted, removing them may take a number of years.

A third general approach would be to recycle some or all of the increased oil revenues that will flow to the federal government and oil companies back to consumers and perhaps to state government agencies. Recycling is designed to cope with both macroeconomic and equity dislocations arising from oil supply interruptions by rapidly restoring purchasing power. Although the normal budgetary processes ultimately recycle revenues, the short-term income loss to consumers not only creates a large drop in effective demand for goods and services but also causes severe hardships for the poor.

Legislation introduced by Senators Bradley and Percy would provide a system to recycle additional revenues collected from the Windfall Profits Tax and the corporate income tax during an interruption, as well as from special taxes and levies imposed at that time. This legislation would recover the oil revenue transfers to the federal government, and perhaps some to the oil industry, distributing them quickly to individuals. It would represent one element of a mildly expansionary fiscal policy and a method for dealing with some of the income losses that would arise from an oil price shock (U.S. Senate 1981). Depending on monetary policy and wage settlements, this form of recycling could give the economy a useful stimulus or it could help overheat it. It also could distribute income progressively.

Three arguments are often used in opposition to recycling. First, the macroeconomic benefits are open to question. Deficits will rise during supply interruptions, because revenues will fall from declining income and because expenditures for government transfer programs, such as unemployment insurance, will increase. Second, some fear that potential government inefficiency might actually exacerbate the drag on the economy by failing to rebate checks quickly and efficiently. Finally, some argue that any federal intervention in the marketplace is a bad thing.

The last two arguments are far from fatal. Recycling systems can be designed to get money out quickly, assuming a willingness to live with a system that is not perfectly equitable or completely immune from fraud. An ideological argument against all intervention could result in greater government meddling. The macroeconomic objection, on the other hand, is much more substantive. Chapters 5 and 6 present two points of view.

Since it is virtually impossible to determine whether the demand reduction shock is more detrimental than the inflationary impacts from higher deficits, it is useful to evaluate whether recycling is more robust than the status quo. Let us assume that a 50 percent probability exists that recycling would do more good than harm, assisting both macroeconomic and equity problems during an interruption. If no recycling program has been developed, then potential benefits under the favorable scenario could never be achieved. If the program were in place but had the effect of creating excessive inflationary pressures, fiscal or monetary policy could be adjusted to correct for the excessive stimulus, including even reducing the size of rebates. Fiscal and monetary policies are always difficult to manage during an oil supply interruption. Because recycling increases flexibility, it appears to have advantages for managing economic policy during periods of extreme stress.

But the two most compelling arguments—which are intertwined—are the political and equity characteristics of recycling. It is naïve to believe that Congress can idly stand by as billions of dollars of revenues are transferred from consumers to oil companies, the government, and OPEC. If price controls are to be avoided, then the government must have a way to prevent, or at least reduce, large wealth transfers. At a minimum, some program to reduce hardships on the poor and elderly will be virtually a political necessity.

Recycling generally makes good economic and political sense. It can restore effective demand during the course of a supply interruption to compensate for demand contraction and it results in a more equitable distribution of income during an oil price shock. If the demand stimulus is too great, compensatory fiscal actions could be taken. The other two options are not nearly as promising. Relying only on the marketplace will create serious hardships for many and will risk reimposition of price controls. Government allocations and price controls are by far the worse option, resulting in shortages during the interruption, artificially high oil consumption, and eventually, large price increases and greater losses in economic growth.

Recycling Issues

Often, the discussion about recycling ends here. Many observers who oppose price controls assume some form of recycling is a political sine qua non for using the marketplace, and many believe it would be necessary to cope with macroeconomic problems. But the debate seldom moves to the detailed policy and implementation issues. Resolution of these issues will be necessary before one can talk realistically about using a market allocation system.

Deciding how much to rebate is the first issue that must be faced in designing a recycling program. Under the Bradley–Percy bill, oil revenues from the current Windfall Profits Tax and the corporate income tax, at a minimum, would be rebated. The bill would also allow for increasing the Windfall Profits Tax or for using sales of the Strategic Petroleum Reserve (SPR) as additional sources of revenue for rebates. Hence, a large percentage of increased domestic oil prices would be rebated, with the potential for providing an even larger level of recycling (U.S. Senate 1981).

A more limited program could tap revenues from the Windfall Profits Tax only, providing recycling equal to 25 percent of the oil price increase. The rest of the increased oil revenues would be transferred to the Treasury in the form of higher corporate taxes or to oil companies and lease holders in the form of higher profits. If it were designed specifically to help the poor, then this more limited source of revenues may be appropriate. If the recycling program were designed to restore a substantial amount of lost purchasing power, however, then some combination of Windfall Profits Tax and corporate income tax revenues would be necessary. Hence, the decision on basic policy objectives would dictate whether to use the formula in the Bradley–Percy bill or choose a more limited revenue base.

In addition, natural gas revenue recycling must be considered for the future. Once natural gas prices become more directly linked to oil prices, either from full decontrol or from partial decontrol in 1985 under the Natural Gas Policy Act, natural gas prices will move up during oil price shocks. Policymakers will need to decide whether revenue from natural gas price increases should also be rebated. If gas revenues are included from a fully decontrolled market, they could equal up to 50 percent or more of the oil revenues available for recycling.

The second implementation issue, the rebate mechanism, is the most difficult politically. If program revenues are only adequate to help the poor, the current eligibility systems for allocating funds to the poor and the elderly could presumably be adapted to this purpose. Or funds could be allocated to the states as part of an expanded Low Income Energy Assistance Program. But should much larger sums be available, such as those suggested in the Bradley–Percy bill, then new distributional formulas must be created. Whenever disbursement of billions of dollars becomes necessary, even if only for emergency conditions, politicians take notice. Hence, it is important to establish the unifying rationale for distribution in advance, before the process itself becomes nothing but a measure of political power.

The three most obvious rebate schemes are: transferring the task to the states, which then become the major source for distribution to consumers; creating a formula that somehow relates rebates to need; and establishing an eligibility formula that provides equal payments to each person or household. Each of these alternatives has its own political constituency, administrative problems, and equity characteristics.

Rebating all of the funds to the states sounds like a political "free lunch," absolving the federal government of the administrative problems of recycling and allowing governments closest to the people to decide how funds would be allocated. Indeed, if funds were only designed to help the poor, then distribution by states may be attractive since they already administer the Low Income Energy Assistance Program. But once the decision is made to pursue a broader recycling program, the attractiveness of state administration fades quickly. It is by no means clear that states are the best implementors of a program designed to cope with national macroeconomic and equity problems. Rebating billions of dollars through fifty disparate systems may turn out to be an administrative nightmare, causing hardships for the poor and delays in stimulating demand. Many states have virtually no machinery to operate such programs. Nevertheless, the idea has sufficient political appeal that it should be reviewed as an alternative. Even if the states are not the proper administrative vehicle for the entire program, it would make sense for states to distribute some portion of the funds—maybe 5 to 10 percent—allowing them to handle particular regional hardships and public facility needs.

Establishing a formula that somehow distributes rebates on the basis of a determination of need would become a political morass. Technically, formulas could be created that provide citizens in different states different rebate levels, based on climate, distance, and other factors that affect fuel use. It is, however, hard to conceive of a formula that would be perceived as being fair to all interests. Once started down the path of relating rebates to need, many politicians will find some need factor peculiar to their own constituencies that deserves special attention.

Some criterion for providing consistent rebates—such as per capita or per household—is desirable. Such a rebate system would be generally progressive, since the uniform rebate would represent a much larger portion of the income of the poor. It would be perceived as fair, since citizens in different states or localities would not receive different amounts. Congress has already shown a willingness to provide funds on such a basis in the Tax Reduction Act of 1975 and House action on the 1977 Crude Oil Equalization Tax rebate. And it would be much simpler to administer, since limited staff would not have to calculate and recalculate data to determine individual eligibility for the credit.

Assuming that decisions have been made on how much to rebate and who would be eligible for the rebates, the third question is implementation. Usually, those in favor of a rebate initially advocate

use of the income tax withholding system as the chief vehicle for recycling funds. No other system covers so many Americans, and the
administrative costs of using this existing apparatus would be relatively low. Chapter 7 advocates using income tax withholding as the
major distributor of rebates. At first, the apparent advantages of such
a system seem to overshadow the not so obvious drawbacks. But
looking more closely, one finds that millions of Americans, particularly the elderly, do not have taxes withheld. Hundreds of thousands
of small businesses would have to change withholding levels, perhaps
two or three additional times during a year. If other programs are
also used to dispense rebates, such as the Social Security system and
welfare programs, then no *one* federal agency bears responsibility
for the overall success of the recycling endeavor.

If this system faces all of these inherent problems, then what kind
of a system might be workable? First, as a political and equity requirement, it must have comprehensive coverage so that as many
people as possible can receive the rebates without creating an extensive and complicated registration system. Second, one agency must
be in charge of the program to ensure uniformity in administration.
Third, costs must be low both to the federal government and to the
private sector. Fourth, it must have minimal potential for large-scale
fraud and abuse. Finally, it must require a minimum of congressional
action when the country faces an actual oil shortage. One system
designed to meet these criteria is described in Chapter 8.

One of the most important criteria in choosing among alternative
recycling systems is the capacity to be set up in advance. This concern arises from the premise that Congress will be most capable of
developing a rational system when the threat of oil interruptions is
small. During such relaxed periods the Congress can act without
immense constituent pressure, allowing time for the congressional
leadership and the administration to work together in overcoming
legislative bottlenecks.

During the actual shortage, however, congressional leaders are
beset with constituent fears and the general confusion surrounding
supply interruptions. Congressional action on President Carter's proposed gasoline rationing plan shows how those fears and concerns
affect legislators. During the summer of 1979, the Carter administration attempted to gain congressional approval for its gasoline
rationing plan under the procedure established under the Energy Policy and Conservation Act of 1975. After the administration was
forced to agree to three amendments to the plan in the Senate, that

plan was soundly defeated in the House. A majority of House members felt that the rationing plan would not be fair to its constituents. After the summer gasoline shortages were over, a slightly different legislative vehicle was created and the gasoline rationing plan was approved.

Decisions on preimplementation represent the fourth and final issue. Preferably, the formula for distribution should be authorized in advance. For example, legislation might specify that rebates would be made to all citizens on a per capita basis from increased oil revenues from the Windfall Profits Tax and the corporate income tax. The secretary of the Treasury would estimate these revenues and monitor their outflow during the interruption. Under this system, no congressional action would be necessary during the oil cutoff, since the beneficiaries and source of revenues would both have been preauthorized.

If this form of complete preauthorization were politically unacceptable, the authorizing legislation could predetermine everything except for appropriations. Congress would need to appropriate funds at the time of an interruption, based on estimates of increased tax revenues accruing from the higher world oil prices. By design, the appropriation should be relatively pro forma, but during the politically charged period of an interruption, the temptation to play with the distribution formula could be great.

A system that required significant congressional action during an interruption would be the most unworkable. If reduced tax withholding were the major vehicle for recycling, the Senate Finance and House Ways and Means Committees would need to change withholding rates during the interruption and other committees would need to authorize and appropriate funds for other recipients. This system would require an almost unprecedented amount of congressional coordination.

Progress is finally being made on the implementation issues posed by recycling. Alternative administrative approaches have recently been presented and debated. The next step is to select one or more options and subject them to rigorous scrutiny.

STOCKPILES

Although almost all energy experts and politicians agree that a strategic petroleum reserve is desirable, serious disagreement exists on the purpose of the reserve and how it should be managed. Some view it as a tool of military and foreign policy, only to be used during periods of dire national security threats. The majority of political leaders and interested citizens view the reserve as a source of fuel to prevent physical shortages. The persistent concern of New Englanders and Hawaiians over regional reserves, for example, attests to the depth of this feeling. Finally, most economists and energy experts view the reserve primarily as a tool to minimize price increases during and after oil interruptions.

The SPR can serve all of these goals. Considering the serious economic damage inflicted by past supply interruptions, it would be foolish *not* to use the reserve to mitigate economic damage. If the interruption became extremely serious, the SPR could be used to help the United States meet its obligations under the International Energy Program of the International Energy Agency (IEA). Petroleum obviously would be available for military purposes and for purposes of maintaining the economy in the face of implicit or explicit political blackmail.

Understanding the relationship between these goals is a necessary first step in establishing priorities for the SPR management system. While oil supplies for military operations and for essential security functions must represent the highest priority, the amounts required are not large unless defense needs are defined as mobilization for a large and protracted world conflict. Unless the shortage is extremely large or for political reasons the market is not allowed to work, the SPR would not be needed to offset physical shortages. The major purpose of the SPR is to reduce the price impacts from an oil supply interruption. SPR operations should be geared toward having a maximum impact on reducing world oil price pressures, while still ensuring adequate flexibility to cope with actual shortages or even military needs should the interruption become grave.

Price increases will be driven only partly by the size of the shortfall itself. Inventory accumulation, based on expectations of higher prices in the future, can lead to sharp spot market price increases which, in turn, lead to higher contract and official prices. Hence, a major operational goal of the SPR is to affect private stockpiling decisions, both in the United States and abroad, by dampening spot market price increases and changing expectations about future prices. Another goal is to keep adequate reserves against the possibility of actual shortages, particularly shortages that might affect national security.

During the early stages of an interruption, the primary goals should be to minimize spot market price increases and to affect perceptions about future prices. Either petroleum must be released quickly from the SPR or the industry must perceive that the SPR would be used decisively if prices rise above a certain level. At the same time, decisionmakers must have working rules of thumb on how much oil should be held back for more serious interruptions. Empirical work (see Chapters 10 and 11) gives us a better idea about what levels of shortage and national security stocks are required. Necessarily, decisions will be heavily dependent on the amount of SPR oil available at the time and the nature of the particular interruption.

Some observers believe that excessive concern over national security could inhibit decisionmakers from ever using the strategic reserve. They point to other strategic stockpiles that were never drawn down and argue that the SPR might suffer the same fate. But there are strong countervailing arguments. Strategic materials such as chromite, cobalt, tungsten, manganese, and tin represent small components of GNP, while the cost of crude oil alone currently equals about 5 percent of the U.S. GNP. The experience from the 1979 interruption, with substantial GNP losses in most of the industrialized countries, should discourage policymakers' holding back too long on stockpile use. The political pressures during an interruption will force politicians to do something, or at least appear to be doing something, and withdrawal of the SPR represents a relatively painless step. Decisions may well be deferred as policymakers try to keep their options open as long as possible, but, faced with serious economic problems and political turmoil, it is highly likely that the reserve will be used.

The size of the reserve at the time of the interruption will have a significant impact on decisionmakers' willingness to use it. If the reserve were filled to the 750 million barrel target, it could be used comfortably in interruptions comparable to those that have already occurred. If, for example, the United States had decided to cover its proportionate share of the 2 mbd Iranian interruption, about 700 thousand barrels, it could have withdrawn oil at this rate for about a year while depleting only 255 million barrels of storage. If an interruption had resulted in a U.S. withdrawal of 2 mbd, however, half the reserve would have been used up in six months.

Several studies have suggested that an extremely large SPR (one to two billion barrels, and under a few assumptions, more) is warranted

(Plummer 1982b; Rowen and Weyant 1982; Hogan 1982; U.S. Department of Energy 1980). Most of these estimates assumed larger import levels and a more menacing security environment; if they used more current assumptions, optimum size would be lower, although higher than the current 750 million barrel target (U.S. Department of Energy 1982).

Although the ultimate size of the reserve is not a trivial problem, other pressing issues face policymakers:

- the fill schedule for the reserve
- incentives for private stockpiling
- the trigger for using the reserve
- the mechanism for selling oil from the reserve
- criteria for decisionmaking during an interruption
- the trigger for phasing out withdrawals

The urgency given to filling the strategic reserve must be considered in the context of the level and prospects for total U.S. buffer stocks. The strategic reserve increased rapidly during 1981, averaging 335.8 thousand barrels per day. Prior to the deferral of funds for Phase III storage facilities, projected expansion of the reserve had implied a fill rate of 189 thousand barrels per day from 1982 to 1989; but with the proposed deferral, the fill rate had been projected by the General Accounting Office to average closer to 168 thousand barrels of oil per day during fiscal years 1982 through 1990. This average would have only been about half of the 1981 fill rate (U.S. General Accounting Office 1981: 19).

The Congress established a goal of filling the reserve at a level of 300,000 barrels per day and an absolute minimum at 220,000 barrels per day. The Department of Energy will comply with the 220,000 minimum level for fiscal year 1983, but the fill level will fall in fiscal year 1984.

Constraints on the SPR program are compounded by smaller private stockpiles. U.S. private stocks reached their highest level—1.36 billion barrels—in August 1980. By August 1982, these private buffer stocks fell to 1.14 billion barrels, over a 200 million barrel difference. Imports fell only slightly over this two-year period, from 5.7 to 5.2 mbd. The SPR grew from 91 to 267 million barrels over the same period, an increase of 174 million barrels. Hence, despite the actions by the Reagan administration in filling the strategic reserve, the United States actually lost buffer stocks during this period.

If the United States were to fill the reserve again at 300 thousand barrels per day—a level it achieved in both 1978 and 1981—then it would need to lease above-ground storage or tankers to store oil, or lease the oil itself. It appears that leasing, at least in the short term, is cheaper than storage in the salt domes. The minimum fill level of 220,000 barrels per day would not require leasing for fiscal year 1983 but might require some storage in fiscal year 1984.

Some have suggested that new tax credits be established to encourage industry to hold larger private stocks. Creating incentives for the private sector to hold greater stocks is attractive because government involvement could be minimized and because stock buildup would not be tied to the schedule for leaching salt caverns. Unfortunately, creation of a workable tax credit is no less complex than a regulatory program. A tax credit for the more than one billion barrels of industry stocks would be extraordinarily expensive, providing large windfalls to oil companies for holding stocks they would have to hold anyway. To overcome this problem, a base would need to be created to determine which stockpiled oil represented strategic insurance and hence should be eligible for the credit. But deciding upon a base period is not simple. Should it be the level of stocks oil companies held before the 1979 interruption or should it be set at a higher level, reflecting experience throughout 1979–81? If the base is set too high, the incentive will be too weak to encourage much stockholding; if set too low, large windfalls will accrue to oil companies. It would take extraordinary judgment to come even close to developing a cost–effective incentive, considering the different stock situations of oil companies.

A joint public–private corporation, which would store oil and manage drawdowns during an interruption, represents another, radically different option. Such a corporation would own and be responsible for managing between one-fourth and one-third of the emergency reserves, purchased with government-guaranteed loans or with a portion of the SPR as collateral. Industry participants would pay storage fees to the corporation but would put up none of their own capital. Corporate management would consist of both industry representatives and presidential appointees.

By relying upon industry expertise and at least partially insulating corporate decisions from politics, many of the management pitfalls of a centralized reserve could be avoided. The corporation's stocks would constitute a flexible reserve at the outset of the interruption, which could be used to quell panic and blunt initial price increases as well as to ensure access to crude oil for the most directly affected refineries. The SPR would be available for use during large interruptions. The corporation option offers an interesting way to coordinate public and private actions during an interruption and to depoliticize initial stockpile withdrawal decisions. So far, this option has attracted little interest.

The next implementation issue revolves around how to begin discharging oil from the reserve. This decision will be difficult because it signals serious concern about the political events precipitating the crisis and about the expected impacts of the interruption on the economy. Two basic theories have been suggested regarding the general approach for discharging the strategic reserve. Some advocate that oil be made available on a predictable basis at the beginning of the crisis to reduce market uncertainty. Others advocate just the opposite, arguing that predictable government actions will skew private behavior, perhaps discouraging oil companies from building their own inventories.

One variant of the second general approach would be to announce that the reserve would be used only during serious interruptions. Under such a policy, refineries would be forewarned that failure to build up some strategic stocks would leave them vulnerable during the early stages of an oil supply interruption. Unfortunately, there is no reason to believe refineries will hold substantial strategic stocks, even in the face of such a government threat. Except for the period following the Iranian Revolution, crude oil stocks have consistently stayed between nineteen and twenty-four days of refinery input—a level necessary for operating the production and distribution system. These levels were maintained in the absence of a strategic reserve; it is hard to believe that higher levels of stocks would be held in the presence of such a reserve.

Considering the cost of holding stocks and the probability that government would not hold to its threat, most refiners are likely to be holding minimal inventories when the next interruption actually strikes. Hence, no apparent advantage flows from a system that shuns early government action in drawing down the strategic reserve. On the other hand, uncertainty about government actions at the onset of an interruption could create a certain amount of anxiety in the world oil market, which, coupled with normal pressures to stock up early during an interruption, could lead to spiraling spot market prices.

If the SPR is designed to pursue the operational goal of affecting private stockpiling decisions, then it must be released quickly and predictably during an emergency. Although one can always hope that future decisionmakers will have rare prescience and know exactly how quickly to release oil, most likely some form of releasing trigger will need to be developed in advance.

One possibility for such a trigger would be to relate SPR withdrawals to some estimate of the world oil shortfall. For example, during the first quarter of 1979, the net 2 mbd world shortfall resulted in about a 700 thousand barrel per day U.S. shortfall. The United States could decide to trigger the SPR automatically based on the U.S. share of a particular level of world shortage. The problem with the quantitative supply trigger is that data are poor during an interruption and serious disagreement could exist as to the magnitude of the shortage.

Another possible trigger could be based on price changes. For example, the United States could agree to discharge so many thousand barrels of oil per day once the spot price exceeded the official price by a specified amount. The price trigger has the advantage of empirical verification since a number of publications continually quote spot market prices.

After deciding the basis for triggering, a decision must be made on the method of release. The options range from distributing the SPR through competitive sales to various forms of government allocations. As an example of the latter, SPR oil could be made available at preferential prices to refiners whose supplies were cut off. If SPR oil is to be allocated on the basis of some form of market system, the question then arises as to which system or systems would be best. Competitive bidding represents the most commonly discussed market form for releasing the SPR, primarily because it most nearly mirrors the spot market.

The operational goal of influencing expectations about future prices does not require immediate physical release of oil, however. Indeed, systems that provide contractual guarantees that oil would be delivered in the future may affect future price expectations more than a direct discharge. In Chapter 10, Shanta Devarajan and Glenn

Hubbard conclude that world oil prices would reach lower levels if a futures market were employed rather than direct sales.

Futures and options markets affect expectations by ensuring that supplies will be made available in the future and thus allow refiners and major oil consumers to substitute financial obligations for actual oil purchases in the spot market. The strategic reserve could initially be integrated into one of these financial markets by deciding what quantities of oil would be available through forward contracting mechanisms and then actually making them available on the open market. If the futures market were used, contracts would be sold for delivery of a specified number of barrels of oil for different time periods. If the options market were used, the government would offer for sale options to purchase x barrels of oil at y dollars per barrel at different times in the future. The oil would be made available to the holder of these financial obligations at the end of the option period or the termination of the futures contract. Should the spot market price fail to rise to the futures price, then either the option would not be exercised by the holder or the federal government could choose to purchase back the futures contract.

A futures or options market would have many advantages for government decisionmakers. They could initiate such market transactions without making any declaration about the nature of the political event that precipitated an oil interruption or about the potential for a more serious shortfall and higher prices. Release of SPR oil, on the other hand, might be considered tantamount to U.S. government recognition of an impending crisis. By allowing the market, rather than the government, to determine the future value of oil, the government might be able to attain greater flexibility during the early stages of an interruption. If options bids are so high that they give policymakers cause for concern, they can reopen the bidding with a greater quantity of oil, thereby reducing the options prices. Market prices for futures contracts, however, could not greatly exceed spot prices since firms always have the option to buy in the spot market for future use or speculation. Finally, under both of these alternatives, a decisionmaker would know that rights to oil sold on such markets could be repurchased in a dire emergency, such as war.

A futures or options market provides refiners with an alternative to stockpiling physical quantities of oil. By providing such an option, these alternatives might have the detrimental effect of reducing stock levels prior to an interruption. But during an interruption, they would have two distinct advantages. First, refiners would know that oil supplies would be available to them in the future, allowing them to draw down their own stocks. If direct SPR sales are used, however, firms would be hesitant to use stocks because they could not be certain of future availability of supplies. This uncertainty over supplies could result in more demand being placed on the spot market and could thus raise world oil prices. Second, a futures market would allow greater certainty and flexibility. In essence, a futures or options market might slightly reduce stock levels going into an interruption but would allow stocks to be used rather than hoarded during the course of the interruption. Finally, a futures or options market could help address the concern that certain independent refiners are unable to acquire crude oil by providing an alternative to the spot market.

This approach is designed to influence perceptions of futures prices and, hence, stockpiling decisions. It would be particularly effective in dealing with an interruption similar to the Iranian episode, where the direct shortage was equal to less than half of the extra stockpiling that took place over a two-year period. Conditions could exist, however, in which certain refiners need physical supplies immediately, either because their supplies were directly cut off from the interrupted source or because the total shortage is so great. Under these conditions, direct sales of oil could augment the futures or options market alternative.

Not enough is known to conclude that the futures and options markets are viable. Crude oil futures market trading on the New York Mercantile Exchange and the Chicago Board of Trade has only recently begun. Congress may have a hard time swallowing the idea that instruments normally used for speculation could be a vehicle for distributing SPR oil. The creation of such a market would presumably diminish incentives for stockpiling, although any predictable method of releasing SPR oil would have a similar effect. Despite the unknowns and potential drawbacks, the potential for use of these markets, as a means of both affecting expectations and overcoming barriers to SPR withdrawal, make them potentially strong contenders to supplement direct distribution of SPR oil.

Many of the management questions posed can be decided before an interruption actually strikes. The determination of size and configuration of the reserve can be made during noninterrupted periods.

Even triggers for releasing oil can be created before the actual interruption occurs.

But not all decisions can be made in advance. During the course of the interruption, difficult choices would still be necessary on balancing efforts to minimize current price increases against assurance of future supplies should the situation deteriorate. These decisions cannot be codified in contingency planning manuals, as the ultimate decisions will be heavily dependent on decisionmakers' perceptions about the future.

Strategies can be developed in advance, however, to assist policymakers during the course of an actual interruption. A number of steps should be followed in developing these strategies. First, simple rules of thumb should be developed. One possibility might include, in addition to the price trigger, a decision that the United States would release the SPR in quantities equivalent to its share of the shortage during small interruptions, release some smaller percentage during larger interruptions, and automatically reduce the rate of SPR withdrawal when the reserve falls to certain predetermined levels. Development of these rules must engage top government officials — those at the deputy assistant secretary and assistant secretary levels. There must be a realistic recognition that these rules of thumb would not be slavishly followed during an interruption but would provide a benchmark for decisionmaking. This limited form of contingency planning can be useful to prevent ad hoc decisionmaking during a period of constant media and congressional bombardments.

These general strategies must be accompanied by careful planning if emergency measures are to succeed. Detailed procedures would be necessary to conduct a competitive auction or to initiate futures or options sales. The use of such markets raises a host of legal, institutional, and implementation issues. If emergency conservation and fuel-switching programs were part of a response strategy, more detailed work would be required. The recycling proposal requires substantial setup effort. Before much of this work can be undertaken, basic decisions on strategy are required.

Deciding how to gracefully phase out SPR releases is the last issue. Since uncertainty will remain high for some time after the actual interruption of supplies subsides, premature withdrawal could result in another surge in spot prices. For example, the Saudi Arabian cutback announced in April 1979 helped send spot prices soaring, showing how a modest change during a disturbed market can adversely affect expectations. Looking at the 1979 experience it is difficult to determine when SPR withdrawal could have occurred gracefully until the market finally calmed down in early 1981. Before reducing SPR withdrawals, decisionmakers will want to see favorable trends in terms of disrupted supplies coming back on line, demand falling, spot prices falling toward contract prices, and inventories stabilizing. The pressures to terminate SPR withdrawal prematurely may well be substantial.

The SPR—the only energy emergency tool that currently has widespread political acceptance—is plagued with a lack of agreement on purposes, and inadequate implementation measures. Although progress on filling the reserve during 1981 was impressive and the current implementation efforts are extensive and point in the right direction, there are too many unresolved issues left for a policy tool of this importance.

INTERNATIONAL COOPERATION

Up to this point we have discussed use of the SPR as though U.S. withdrawal decisions could be made in a domestic vacuum. Yet, U.S. stocks represent only 25 percent of free world stocks and 40 percent of OECD stocks (U.S. Department of Energy 1982b: 98). In fact, U.S. actions to affect spot market prices and expectations about future prices will depend heavily on the actions of other allied governments and companies in those countries. When the actions of other governments and countries complement those of the United States and vice versa, the benefits can be substantial. Hogan (1982) estimates potential benefits of $70 billion from reducing economic impacts from an interruption through use of an oil stockpile, compared to benefits of only a few billion dollars for attempting to avoid cooperation. In Chapter 11, Glenn Hubbard and Robert Weiner show substantial decreases in world oil prices from coordinated drawdown. In Chapter 2, Dan Badger argues that much of the price shock of the 1979 Iranian episode could have been avoided with a calculated drawdown of stocks under the control of IEA member governments.

History provides little evidence that cooperation will be easy to achieve. During the 1973 Arab oil embargo the OECD countries were often at odds. During the 1975–78 period, most Western nations pursued independent courses. France, Italy, and the United Kingdom sought bilateral deals with the producers in the form of government agreements and exchanges of oil for technology. The United States developed special relationships with Saudi Arabia and Iran.

In May 1979, the United States provided a $5 subsidy for imported middle distillates to attract supplies supposedly being diverted to Europe, making this subsidy available through the price control program. Japan, cut off from its third-party supplier relationships with major oil companies, aggressively entered the spot market to acquire supplies.

The price runup following the Iranian Revolution had a chilling impact on many of the Western nations, inducing them to seek new forms of cooperation. At first, they agreed to import targets at the Tokyo Summit, then discussed further implementation steps at the Venice Summit. When demand for imported oil fell dramatically in 1981, the quotas were quietly shelved.

No system exists to draw down stocks apart from the IEA sharing agreement. Under the agreement, stock withdrawal is assumed to meet any country's shortfall that is not met by conservation. The ninety-day stock requirement is not aimed at reducing price impacts from an interruption, does not contain provisions for cooperation among the participating countries, and makes no provision whatsoever for drawing down stocks during interruptions of 7 percent or less.

After the experience of the Iranian shortfall, the IEA staff developed a stocks paper that suggested the desirability of coordinated management of a small amount of stocks during an oil interruption. In Chapter 4, Dr. Lantzke, Executive Director of IEA, describes this effort. Another proposal under discussion among Western governments was developed by BP (see Chapter 4). The BP proposal is an interesting idea for oil company self-insurance, which, under certain limited conditions, might discourage a surge of spot market prices. It also has the advantage of not requiring either government money or government decisions to operate successfully. But this five-day reserve, equal to about 175 million barrels, would do little to stem the tide of an Iranian-type oil shock, where private stocks increased 700 million barrels above normal levels. With ample overall spare capacity, the BP proposal could provide individual oil companies with flexibility should they encounter delays or problems from a certain supplier. Hence, the BP proposal could represent a useful mechanism for oil companies but would provide little insurance for any inter-ruption that raised expectations of further turmoil and higher prices.

The IEA and BP proposals both reflect a need for cooperation in discharging stocks during subtrigger interruptions. Both are geared to current political realities, taking into account the resistance by some governments to allied cooperation on stock withdrawal. Consequent-ly, both are quite modest. Considering the direction in which govern-ments are moving, it is useful to begin thinking about a more ambi-tious proposal. The IEA sharing agreement expires in 1984, providing an opportunity to shift the focus of cooperation among consuming nations from sharing shortages to sharing stocks.

CONCLUSION

The 1979 oil supply interruption spawned a new cottage industry of energy security conferences and diagnostic research. These efforts have given us a good understanding of what happened during the 1979 interruption, when a small shortage was translated into exten-sive economic damage. They have clearly indicated which public policy measures can actually exacerbate shortages. Unfortunately, little agreement has been reached on what should be done to cope with future interruptions. The only clear-cut recommendation, at least from the experts, is that stockpiles are a good idea and that government intervention during the last interruption made the problem worse.

Paying obeisance to the superiority of the marketplace for distrib-uting shortages still begs the question of the serious macroeconomic dislocations and personal hardships that arise from oil supply inter-ruptions. Some argue that recycling would be necessary to deal with both problems; others argue that recycling might stimulate more in-flation or that revenues would not be distributed fast enough to over-come the oil price drag on the economy. Most experts would agree that something must be done to cushion the blows of an oil price shock on the poor and that recycling could be the political sine qua non for allowing the market to function during an interruption.

Stockpiles may be generally accepted as a good idea, but substan-tial disagreement exists about how big they should be, how fast they should be filled, what their major purpose is, how they would be triggered, and how they should be managed during a crisis. The most pressing question is the fill rate, followed closely by the need to develop mechanisms for the discharge of the oil.

There is little basis for consensus among IEA members on the issue of international cooperation in energy security. Many Americans do not believe the IEA agreement will actually work, and yet it continues to remain a legal commitment among the IEA participants. Many otherwise knowledgeable people are concerned that oil will be shipped from the United States to other countries as a result of the IEA agreement—an eventuality that is unlikely. U.S. policymakers have been wary of schemes to promote cooperation in managing stock withdrawals, even though the United States, which will have the largest reserve, has much to gain. Perhaps much of the problem arises because so little has been written about the benefits of cooperation, particularly the benefits of joint cooperation in stockpile management. Chapter 11 encourages scholars to think about and begin to quantify the benefits of cooperation during oil supply interruptions.

Progress on energy security issues is being made, even if haltingly. Recycling alternatives are finally being debated as real administrative options. U.S. Department of Energy staff have shown an openness to new ideas and approaches and are working on a range of contingency planning issues. There is hope that a management framework could be developed before the next interruption. But there is certainly no assurance.

Today's worldwide spare production capacity gives us a respite that can be used to prepare for future energy emergencies. As a practical matter, however, policymakers are likely to turn to other issues now that the threat is less imminent. Most of the measures discussed in this chapter will take a number of years to move from policy agreement to congressional action (where necessary), to the stage of actual policy implementation. If these emergency preparedness policies and programs are not pursued vigorously, they could easily be postponed until oil markets are again tight—or slip into oblivion. If we could accomplish the policy development and implementation tasks discussed in this chapter during the time that our vulnerability is lowest, we would then be better prepared to face the possibility of tight markets and increasing vulnerability later this decade.

REFERENCES

Deese, David A., and Joseph S. Nye, eds. 1981. *Energy and Security.* Cambridge, Mass.: Ballinger Publishing Company.

Hogan, William W. 1982. "Oil Stockpiling: Help Thy Neighbor." Harvard Energy Security Program Discussion Paper Series, H–82–02.

Krapels, Edward N. 1980. *Oil Crisis Management.* Baltimore: Johns Hopkins University Press.

Ostry, Sylvia. 1981. "A View from the OECD World." Paper presented at the Atlantic Institute Conference: "Conflict or Cooperation in the 1980s?" Brussels, October 22–24.

Plummer, James L., ed. 1982a. *Energy Vulnerability.* Cambridge, Mass.: Ballinger Publishing Company.

Plummer, James L. 1982b. "U.S. Stockpiling Policy." In *Energy Vulnerability,* edited by J. L. Plummer, pp. 115–148. Cambridge, Mass.: Ballinger Publishing Company.

Rowen, Henry S., and John P. Weyant. 1982. "Reducing the Economic Impacts of Oil Supply Interruptions: An International Perspective." *Energy Journal* 3, no. 1 (January): 1–34.

U.S. Department of Energy. 1980. Office of Natural Gas and Integrated Analysis. "The Energy Problem: Costs and Policy Options." Staff Working Paper, March.

_____. 1982a. Assistant Secretary for Environmental Protection, Safety, and Emergency Preparedness. "Report to the President and the Congress on the Size of the Strategic Petroleum Reserve." DOE/EP–0036. Washington, D.C.: U.S. Government Printing Office, May.

_____. 1982b. *Monthly Energy Review.* DOE/EIA–0035 (82/09) (September): 98.

U.S. General Accounting Office. 1981. "Strategic Petroleum Reserve: Substantial Progress Made, But Capacity and Oil Quality Concerns Remain." EMD–82–19. Washington, D.C.: U.S. Government Printing Office, December.

U.S. Senate. 1981. "Emergency Preparedness Act of 1981." (S. 1354). Governmental Affairs Committee. 97th Congress, 1st Session. Washington, D.C.: U.S. Government Printing Office.

Verleger, Philip K. 1982. *Oil Markets in Turmoil.* Cambridge, Mass.: Ballinger Publishing Company.

Yergin, Daniel H., and Martin J. Hillenbrand. 1982. *Global Insecurity.* Boston: Houghton Mifflin.

I THE OIL MARKET

2 THE ANATOMY OF A "MINOR DISRUPTION"
Missed Opportunities

Daniel B. Badger, Jr.

Understanding the past is important in forging policies for the future. While a "Maginot Line" mentality may be momentarily comforting, the danger of failing to understand recent history will eventually become apparent. The 1978–81 period provides an excellent case study of the dynamics of a relatively small interruption, raising fundamental questions about what governments could or should do when faced with oil interruptions. This chapter traces these developments and then analyzes the actions of governments. It concludes that active government intervention could have avoided some of the worst effects of the 1979 supply shortfall.

PRELUDE: 1970–78

When the world oil price leapt from $2 per barrel in 1972 to $11 per barrel in 1974, the immediate result was recession in the industrial economies and an abrupt end to the rapid rise in oil consumption of the 1960s and early 1970s. Between the first quarter of 1970 (1970:1) and the fourth quarter of 1973 (1973:4) free world oil consumption grew at an annual rate of over 6 percent (see Figure 2-1). But from the beginning of 1974 through the end of 1975— the period following the Arab oil embargo—consumption declined at the rate of 2.5 percent annually. Consumption growth resumed at a

33

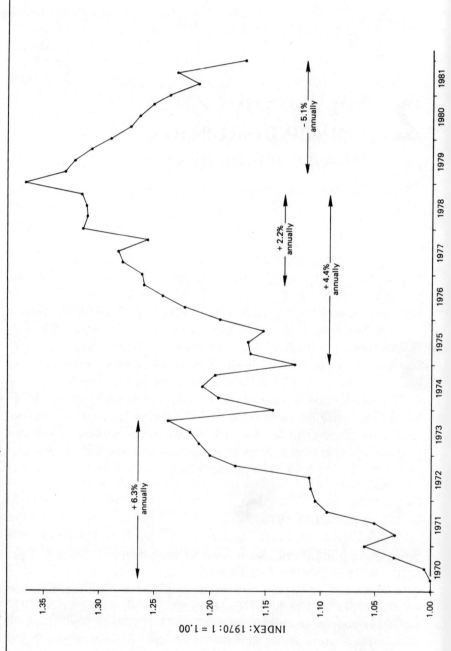

Figure 2-1. Free World Consumption Index.

fast pace at the beginning of 1976, averaging 5 percent annually up to the beginning of the Iranian Revolution in October 1978.

The underlying consumption trend at the beginning of the Iranian Revolution should not, however, be overstated. Much of the rapid increase in 1976 was due to economic recovery. Measured from 1976:4 through 1978:4, consumption grew by slightly more than 2 percent annually. During 1978 itself, the consumption trend was essentially flat. Thus, it seems likely that if there had been no supply disruption in the fall of 1978, consumption might have continued growing at an annual rate of 1 to 2 percent[1] (see Figure 2-1).

On the producers' side, the mid-1970s saw an adjustment to the fivefold increase in revenues tempered by growing concern with the slackness of the market and the steady decline in the real value of revenues. The first perceptible tightening of the world crude oil market since the 1973-74 disruption occurred in the summer of 1978, as companies moved back into the market for supply that, during the first half of 1978, had been taken from the surplus inventories accumulated during the last half of 1977. This inventory drawdown may have marginally overshot its mark since, using 1976 as the norm, overall stock levels on October 1, 1978 were 100 million barrels below "normal operating levels."[2] This difference is, however, well within the margins of estimation error both for norms and for actual levels.

THE IRANIAN DISRUPTION: OIL MARKET CHAOS

The gathering storm of revolution in Iran first affected oil supply when workers went on strike in the oil fields during the second half of October 1978. After a confusing up-and-down period, exports fell to zero during the last week of December and were not resumed until the following March. Figures 2-2 through 2-5 outline, using quarterly averages, the patterns of supply, demand, stock movements, and prices in these and subsequent months.

Supply and Deliveries

Despite the loss of over 2 million barrels of oil per day (mbd) from Iran during 1978:4, there was more than enough spare production capacity to cover the loss. With production increases from other OPEC nations, notably Saudi Arabia, production for the fourth quarter was 1.4 mbd *higher* than in the previous quarter. The total amount of additional production outside of Iran was 3.6 mbd, of which Saudi Arabia accounted for 2.4 mbd (see Figures 2-2 and 2-3).

In 1979:1 with the loss of an additional 3 mbd from Iran, only 1 mbd of spare capacity were available to offset the loss. Production for 1979:1 was 2 mbd less than in the previous quarter. With a strong recovery of Iranian production in 1979:2 (an increase of 2.9 mbd over 1979:1), total free world supply was back at the level of 1978:4. The effect of Iran's increase was partially offset by Saudi Arabia's 1 mbd cutback.

During the remainder of 1979, production remained above the 1979:2 level, although Iran was drifting slowly downward. In 1980:1, the decline in Iran accelerated and, combined with declines by other OPEC producers, set in motion a steady downward trend in world production totals. This trend was, of course, merely a reflection of the consumption trend.

Despite the disruption in Iran, supply in 1978:4 was nearly 2 mbd greater than seasonally adjusted deliveries, implying that stocks were being drawn by this much less than the normal seasonal pattern.[3]

Seasonally adjusted deliveries jumped by 2 mbd in 1979:1. While some portion of this may have been attributable to unusually cold weather in Europe, it is likely that much of this increase represented stock buildup by consumers. The combination of this together with the 2 mbd supply cut reversed the balance of the previous quarter. Thus, in 1979:1, inventories were drawn down by 2 mbd more than the normal seasonal pattern.

Stocks

After drawing down stocks during 1978:4 and 1979:1, industry began a record-breaking stock buildup through the rest of 1979 and most of 1980. At the beginning of the Iranian disruption there was at most a modest deficit (100 million barrels) between actual commercial stock levels and normal operating levels. This deficit increased to the 200 to 300 million barrel range during 1978:4 and 1979:1. Then, in 1979:2, the deficit began to narrow and, by the end of

Figure 2-2.   Free World Oil Supply and Deliveries from Refineries to Consumers.

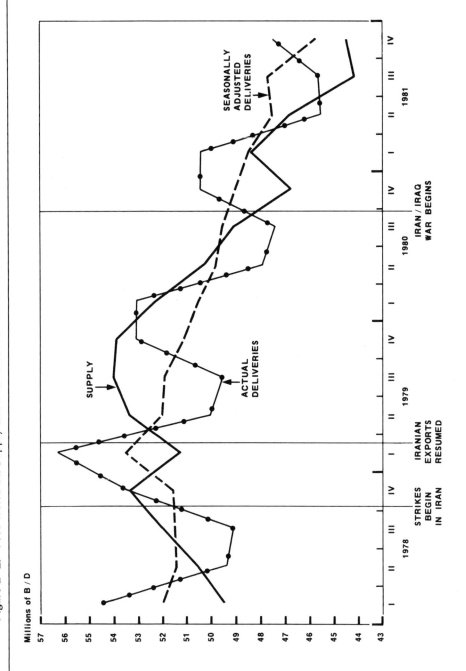

Figure 2–3. Changes in Oil Production Relative to 1978: 1 (mbd).

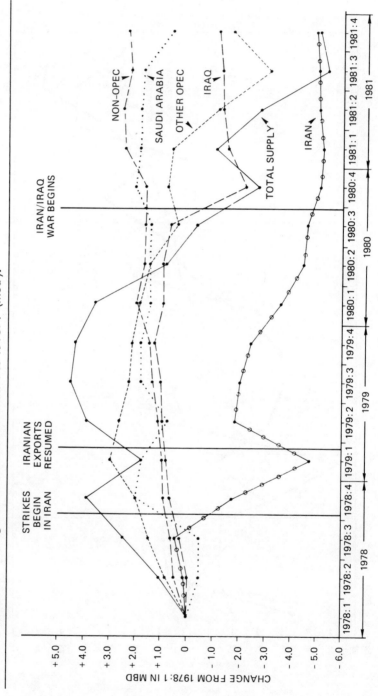

1979:3, actual levels were again above normal operating levels. Due to uncertainty over future supply and the expectation of steadily rising prices, however, the industry's *desired* stock levels were certainly much higher than the normal historical operating levels through this period.[4] There was, therefore, considerable unsatisfied demand for stocks, despite the pattern of rapid stock build.

Between July 1, 1979 and July 1, 1980, the industry built stocks by about 750 million barrels above normal operating levels — an average increase of over 2 mbd. Some of these large excess stocks were worked off during 1980:3, but by the time the Iran–Iraq War broke out, the total commercial cushion above normal operating levels was still about 550 million barrels (see Figure 2–4).

Governments "controlled" an additional 800 million barrels of usable stocks at this point. Governments in the United States, Germany, and Japan owned about 180 million barrels outright. In addition, companies in the European Economic Community (EEC) countries and Japan were holding about 650 million barrels of usable oil (i.e., in excess of minimum operating requirements) which was required to fulfill mandatory stockholding obligations. Governments could in principle have relinquished this oil to company control though they could not necessarily have forced companies to use it.

Because the industry's planners and marketers (and virtually everybody else) consistently overestimated forward demand over this period, actual stock levels at the close of a given month or quarter were generally higher than expected. The significance of this pattern in explaining the stock accumulation in 1979 and 1980 should not be exaggerated, however. While contracts with some producing nations prohibited resales, at least the largest companies retained sufficient flexibility to dispose of unneeded supply through third-party contracts or spot market sales if they so chose. That they did not so choose is a reflection of their concern for supply security and the anticipation of higher crude oil prices in the future. While actual stock levels may have been above expected levels, they were not in excess of the levels industry desired to hold.

The amount of usable oil under government control (700 million barrels) was of the same order of magnitude as the industry's unsatisfied demand for stocks during the first half of 1979 — an actual deficit of 200 million barrels in relation to normal operating levels, plus a desired increment above normal levels of perhaps 400 to 500 million barrels. Therefore, there would appear to have been a good

Figure 2–4. Free World Stocks.

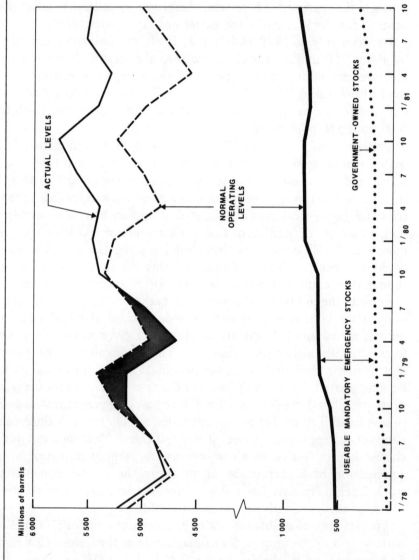

possibility of quelling the chaos that in fact developed in this period by releasing these government-controlled stocks.

Prices

Figure 2-5 summarizes the movement of three price indicators related to Saudi Arabian Light, the OPEC "marker" crude. These include the price for spot products—the so-called "netback cost" that determine crude oil values, the spot crude price, and the official price of marker crudes. These indicators are broadly indicative of overall price movements in the world market. Four price cycles are apparent in the aftermath of the Iranian Revolution.

The first mild cycle of spot price increases peaked in November 1978, when Iranian production was temporarily resumed, but it trailed off quickly. At the peak, spot crude and products were 12 percent and 20 percent, respectively, above their predisruption levels. Official prices did not rise at all.

The second cycle—beginning in late December 1978 when Iranian exports dropped to zero—peaked at the end of February. The peak trailed off when it was announced that Iranian exports would be resumed in March. At the peak, crude and products were 95 percent and 100 percent above predisruption levels. During that period, however, official prices had barely risen.

The third cycle gathered momentum in May and peaked in early June. Crude and product prices rose 180 percent and 140 percent above predisruption levels at this peak. The events that triggered the third cycle are more difficult to pin down. They include Saudi Arabia's second-quarter production cut of 1 mbd, though this was more than matched by a 3 mbd *increase* in Iranian production. Also, Japanese and American refiners tried to replace second-quarter supplies they lost when their third-party contracts with major oil companies were cancelled. Official prices rose during the period from $13.34 to $18.00 per barrel.

A fourth cycle occurred in the fall and early winter of 1979. Iranian production was declining again and production cuts were expected to be made by other producers. The spot crude price peaked 220 percent above the predisruption level in mid-November, and spot product prices peaked in mid-December 150 percent above predisruption levels.

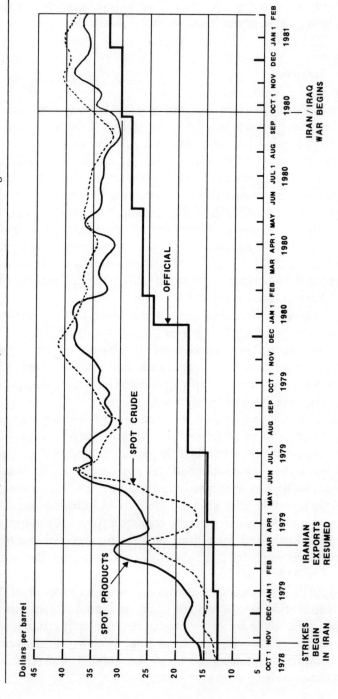

Figure 2-5.   Official Price, Spot Crude Price and Spot Product Value of Arabian Light.

The official price of Saudi Arabian Light moved up in seven steps between January 1979 and the beginning of the Iran–Iraq War. The total increase over this period was 150 percent. Most other official prices moved up faster than Saudi Light, so that Figure 2–4 understates the rate at which average crude oil prices rose. By 1980:1, the official price soared to $24 per barrel. The onset of the Iran–Iraq War led to a comparable rise in spot market prices and increased official prices by $2 per barrel.

ACTIONS BY GOVERNMENTS

Despite voicing great concern, governments took few actions to alter the underlying market forces that were at work during 1979. Many of the actions actually taken had the effect of increasing market pressures. For example, the United States imposed a $5 per barrel subsidy for distillate imports through use of the Price Control and Entitlements Program. This subsidy, announced without warning, irritated other participating governments in the International Energy Agency (IEA) and immediately drove up prices in the spot heating oil market. Ironically, although this action created great consternation, it had virtually no impact on the relative shares of Caribbean heating oil moving to the United States and Europe.

The Japanese government coordinated a program to acquire replacement crude supplies for the various refiners and trading companies that had lost supply from the Iranian Consortium during the summer and fall of 1979. While taking this action to avoid competition among the various Japanese companies, the perception was that Japan was willing to buy crude oil at any price. Most of the major consuming countries took no action to reduce stocks or to discourage high spot purchases; in fact, they implicitly encouraged greater stock building. Only a few governments—Sweden and Belgium, for example—gave permission for companies to draw on portions of the ninety days of emergency reserves required under the EEC agreement.

At the international level, at meetings of the IEA, the Commission of the EEC, and the Tokyo Summit, attention was focused in three areas: targets for demand restraint, targets for import ceilings, and control of the spot market. Short-term action in the first two areas was nonexistent or ineffective. By the end of 1979, the decline in consumption in response to higher prices ensured that both the demand restraint and import ceilings adopted were easily met. Attempts to control spot markets foundered, though more comprehensive reporting systems were established for both crude and product imports in the IEA and the EEC.

CHANGES IN MARKET STRUCTURE

One of the peculiar features of the market response to the Iranian disruption is that supply remained well in balance with consumption levels throughout the period of price chaos in the first half of 1979, and it would most likely have continued to remain in balance for at least another year at predisruption price levels and on the predisruption consumption trend. Why then was the market so concerned about the adequacy of supply?

One answer lies in the industrial countries' sudden attack of anxiety over political instability in the Gulf region. As the Iranian Revolution unfolded during 1979, few considered it altogether implausible that revolution might spread to other oil-producing Gulf states during 1980.

Crude oil access also played a role. The Iranian Revolution led to the abrupt disappearance of the Consortium of major oil companies that had previously been lifting and refining the bulk of Iranian crude production. This Consortium was a significant factor in world oil markets, accounting for some 20 percent of internationally traded crude oil. While some of this crude continued to be produced and sold by the Iranian National Oil Company after the revolution, and while increased production elsewhere more than offset what was lost from Iran, the contractual network that had previously linked refiners to this 20 percent of the international market was dismantled in its entirety during 1979:2. Some of the large companies dependent on the Consortium were able to replace portions of the lost supplies through existing contracts (Exxon in Saudi Arabia, for example). For the most part, however, companies were forced to seek new arrangements to gain access to the supplies which, both inside and outside of Iran, replaced those lost by the Consortium.

Even refiners not dependent on Iran for supplies ran into trouble. Other producing countries cut sales by 10 to 25 percent in some cases, on the pretext of "technical ptoblems." In reality, their use of *force majeure* clauses allowed them to capture much higher prices for crude oil. These supplies were not lost to the market, for they were quickly sold. But they were not sold in the spot market or by contract to new customers at a higher price than was applicable on existing contracts. While total market supplies were not reduced by these changes in market structure, the refiners who lost supply in this way were forced to make new arrangements to replace it.

Those who believe international oil trade is flexible will see no difficulty for the market in coping with these developments. After all, supply was physically available to meet demand during most of 1979. Since the overall supply–demand balance was not affected, price pressure should not have resulted.

Others believe that market imperfections in the form of inadequate information and high transaction costs preclude easy adjustment. They believe that quick rearrangement of access to crude supplies does not take place without friction, helping explain why the market went out of control in 1979: 2 despite an overall supply-consumption balance that was more than adequate. Oil users cut off from supplies often frantically entered spot markets, it is argued, leading to an upward spiral in price. The independent U.S. refiners who were called in by the U.S. Secretary of Energy in the spring of 1979 were not impressed by his staff's demonstration that the supplies lost in Iran had been more than offset by increases elsewhere. The refiners asked the secretary if he could provide telephone numbers to call to buy contract oil.

THE IRAN-IRAQ WAR: THE MARKET IN DECLINE

The key characteristics of the world oil market in September 1980, when Iraqi troops advanced across the Shatt-al-Arab into Iran, included a commercial stock cushion of some 500 million barrels above normal operating levels and a consumption decline trend at a 5 percent annual rate. From Iran and Iraq the combined supply loss in 1980: 4 was 3.4 mbd. This was partially offset by an increase of 1 mbd by other OPEC producers, so that the net supply loss for the fourth quarter was 2.4 mbd. In 1981: 1, 1.6 mbd of this loss was restored, with Iran and Iraq accounting for 1.1 mbd.

Despite the high stocks and declining consumption, spot markets remained excited well into January 1981. The spot crude market reached $40 per barrel in mid-November and again in early January.

As a result of the high spot prices that prevailed when OPEC met in December 1980, the cartel was able to impose a 10 percent price increase. The result was a $2 per barrel increase in the average crude oil price. Product prices peaked in mid-November at 30 percent above their predisruption level. By the middle of January, however, the spot markets went into a steady decline. This dates the end of the crisis.

The Role of Stocks

The high levels of stock available at the beginning of the Iran–Iraq War helped prevent a recurrence of the spiraling price increases that followed the Iranian Revolution. The existence of a large surplus of usable stocks made it less necessary for companies to replace lost supplies in the market in 1980:4 and 1981:1. Even prior to the war, the industry had in 1980:3 begun working off this stock surplus by lifting about 500 thousand barrels per day less oil than required to balance seasonally adjusted consumption (see Figure 2-2). The world stock draw in 1980:4 was abnormally large, reducing total company-owned inventories by 400 million barrels. At the same time, government-owned stocks went up by 20 million barrels— mainly through purchases for the U.S. Strategic Petroleum Reserve. In total, the usable commercial surplus above normal operating levels was reduced by 100 million barrels (from 500 to 400 million barrels), due to the decline in consumption and the reduction in mandatory emergency reserves that took effect at midnight on December 31, 1980 when the base period for measuring the obligation shifted from calendar 1979 consumption to calendar 1980 consumption.

The Role of Governments

Governments' responses to the Iran–Iraq War were considerably more conducive to market stability than was the case during 1979. At the IEA, for example, member governments had agreed within ten days of the beginning of the war to urge their oil companies to use stocks in preference to going to the spot market to cover supply losses. Although this agreement was not difficult to adhere to, since stocks were high and demand was falling, it did represent a nascent effort to manage stockpile withdrawals. So far as anyone can tell, no member government acted domestically to undermine the spirit of this commitment.

A HYPOTHETICAL REPLAY OF 1979-80

With the benefit of hindsight, it is now apparent that the dramatic price increase of 1979 and 1980, while necessary to strike a market-clearing balance between supply and *demand* for oil, was larger than necessary merely to balance supply and consumption. The larger increase was required because demand included not only consumers' demand, but refiners' considerable demand for stocks.

We can illustrate this by working through a hypothetical replay of 1979-80 in which refiners' excess demand for stocks is assumed not to have arisen. We make the following assumptions regarding the consumption and production paths that might then have been observed: (1) consumption rising at 1 to 2 percent annually (consistent with the 1976-78 trend) until the limits of production capacity are reached; and (2) supply at historical levels through 1979, and at higher than historical levels in the first three quarters of 1980.

The rationale for assuming higher than historical production in 1980 is twofold: First, since much, if not all of the historical production decline in 1980 was due to declining consumption, a higher consumption level would presumably have brought forth higher production. Second, since the price path in the hypothetical replay would have been lower, those producers whose production levels are set to satisfy minimum short-term revenue requirements may have produced more. Kuwait, Libya, and even revolutionary Iran might have reduced production less in 1980 if prices had been lower.

Redrawing Figure 2-2 on the basis of these assumptions yields the pattern of supply and consumption shown in Figure 2-6. It is apparent that at the lower growth rate (1%), consumption does not push up against the supply constraint prior to the Iran-Iraq War, while at the higher rate (2%), it reaches the supply constraint just prior to the war. Below we look at the consequences of the Iran-Iraq War in the hypothetical case. Here we simply note that during the period between the Iranian Revolution and the war, a price increase less drastic than the historical one was entirely consistent with the underlying supply and consumption trends.

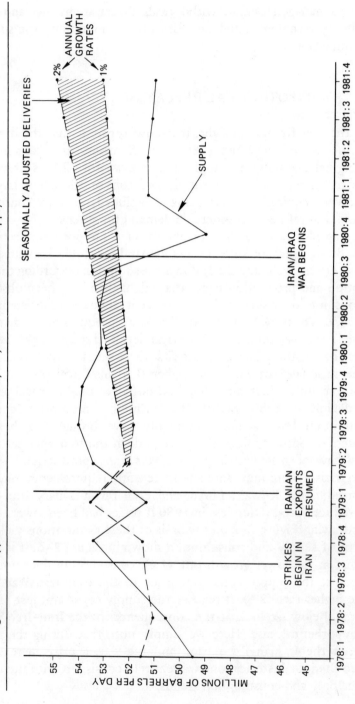

Figure 2-6. Hypothetical Patterns of Seasonally Adjusted Deliveries and Supply.

How Might It Have Been Done?

Many types of government intervention during the first half of 1979 could have had the effect of reducing world oil price increases. These include some relatively *dirigiste* schemes, such as import price ceilings or ceilings on company stock levels, and more "market-oriented" schemes, such as sale of government stocks and relaxation of mandatory stockholding obligations. The essential difference between the two approaches is that the first would attempt to suppress the "excess" demand for stocks while the second would attempt to satisfy it. The second approach has the added advantage that, to the degree it is successful in calming markets, it will also reduce the demand for building up stocks in anticipation of higher prices.

The amount of usable oil under government "control" in the first half of 1979 was of the same order of magnitude as the estimate of unsatisfied demand for stocks. A combination of selling government-owned oil and relaxing companies' mandatory stockholding obligations might well have moderated the market pressure of that unsatisfied demand, leading to lower prices, higher consumption, and higher production.

The Risks

A few pitfalls must, however, be recognized. When Iranian exports were resumed in March 1979, Saudi Arabia cut production by 1 mbd, with the stated intention of "making room" for the renewed Iranian exports.[5] The possibility that the release of government stocks in consuming countries might similarly have caused a matching Saudi production cut cannot be ignored.

In the same vein, it would have been a mistake to allow a slack market to develop in which buyers balked at current prices, causing producers to reduce production. As indicated in Figure 2-6, the assumed level of supply is substantially in excess of annualized deliveries throughout the period 1979:2 to 1980:3, implying a willingness on the part of industry to accumulate stocks. Should the industry have proved unwilling to buy and hold these stocks, it would have been up to governments to step in and keep pressure on the demand side either by buying stocks themselves or by reimposing mandatory requirements on companies.

One must also ask whether a policy that succeeded in moderating the price increases of 1979–80 might at the same time have reduced the economic incentive to produce. This hypothesis seems unlikely in the case of OPEC, since all of the OPEC production increases came in 1978:4 and 1979:1, before contract prices had risen significantly. In the case of Saudi Arabia, which accounted for the bulk of the increase in production in 1978:4, it is clear that the main objective was market stability rather than increased revenue. Increases by other OPEC producers, such as Kuwait, involved a return to production levels that preceded the slack market of 1978 and can be attributed to renewed buyer interest. If anything, the events of 1979–81 have shown that OPEC countries are inclined to produce less at higher prices than at lower prices. The 1979–80 production increases in Alaska and the North Sea were justified by price expectations of the early and mid-1970s. Hence, there is no reason to believe production would have been appreciably less had prices been lower.

Some would argue that government efforts to allow or encourage stock drawdowns would leave consuming countries more vulnerable to future interruptions. This results from the tendency to think of these stocks as being "drawn down," which in turn implies that they have been "burned up."

In fact, when government stocks are sold or mandatory requirements are relaxed, the oil is transferred to the industry. The level of stocks is thereby reduced *only insofar as the transfer causes production to be cut or causes consumption to be higher* than it otherwise would have been. The production cut is, of course, something to be avoided. The higher consumption path is not to be avoided, however, since less drastic price increases and a correspondingly less severe decline in consumption are in fact that objective of using government stocks to minimize price increases and economic shocks.

In Figure 2–7, the historical surplus of usable stocks (including government-controlled stocks) above normal operating levels is shown together with the totals that would have resulted from the hypothetical supply and consumption patterns from Figure 2–6. The stock surplus at the beginning of the Iran–Iraq War is substantially less in the hypothetical positive growth cases than in the historical negative growth case, but even in the higher of the two positive growth cases the surplus is comparable to the historical surplus at the beginning of the Iranian Revolution.

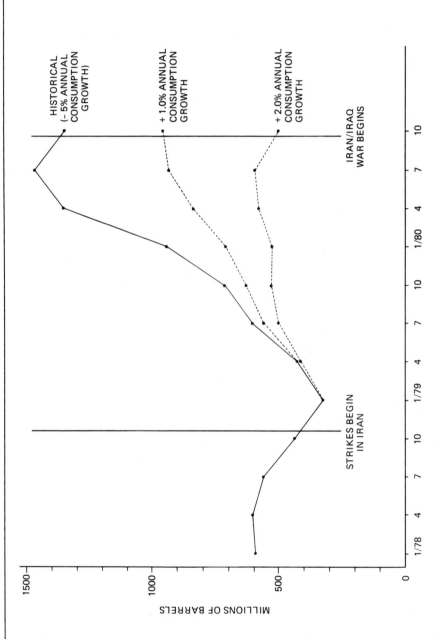

Figure 2–7. Historical and Hypothetical Usable Stock Surplus (*including government-controlled stocks*).

It is clear from Figures 2–6 and 2–7 that under the hypothetical replay, the market would have been in for a very severe supply shock in 1980:4. In the worst case considered—2 percent annual growth in demand—consumption in the period beginning in 1981:1 would have had to be cut back by about 2.5 mbd in order to balance with supply, with about 500 million barrels of surplus stocks available to smooth the transition. At the 1 percent demand growth level, stocks would have been sufficient to moderate prices considerably.

While the oil price increases resulting from the Iran–Iraq War shock would have been severe, two points are worth noting. First, the Organization for Economic Cooperation and Development (OECD) has estimated that in 1980 its member countries lost some 3 percent of gross income, worth about $350 billion, as a result of the oil price increases in 1979–80 and another 4.25 percent, or $570 billion, in 1981. This was part of the cost of having an extra 400 to 800 million barrels of oil on hand when the Iran–Iraq War broke out. One must consider whether this increased security was worth the price.

Second, if in the hypothetical replay the stock surplus at the outbreak of the Iran–Iraq war had been sufficient to allow a smooth transition to the new supply level to be made, the new equilibrium consumption path following the second shock would still have been higher than the historical consumption path. This would correspond to a lower price level and to a savings, therefore, of some portion of the $570 billion that the OECD member countries lost in 1981 as a result of the 1979–80 price increases. On the other hand, if the stock cushion had been perceived by the industry as inadequate to make the transition, additional unsatisfied demand for stocks would have been the result, and the historical pattern of 1979 would have been repeated: Prices would have risen much faster and higher than necessary to balance supply and consumption, and within a year or so, rapidly declining consumption and its attendant economic damage would have resulted.

CONCLUSIONS

The oil price increases of 1979–80 are widely believed to have been the unavoidable result of the loss of supply from Iran following the outbreak of the Iranian Revolution in October 1978. Yet, a closer look at the patterns of supply, consumption, and stocks during this period shows only a relatively small and temporary imbalance between supply and consumption. A surge in demand for stocks appears, therefore, to have contributed strongly to the dramatic rise in market-clearing prices.

A policy of government intervention to release oil to satisfy the excess demand for stocks would very likely have achieved different results. It would have moderated price increases in 1979 and 1980, allowed higher levels of consumption, and still provided surplus stocks to help absorb the shock of the second disruption caused by the Iran–Iraq War.

Some price increase was eventually inevitable due to production capacity limitations either before or as a result of the Iran–Iraq War. It appears likely, however, that if the assumed intervention had been carried out prudently, the total price increase necessary to adjust to the two supply shocks would have been less than the historical increase and would have been achieved in a smoother, more predictable fashion. The benefits of such an intervention would have been measured in terms of hundreds of billions of dollars annually.

NOTES TO CHAPTER 2

1. In relation to the observed historical price increases and consumption decreases, the assumption that demand would have grown at 2 percent annually in the absence of any real price increases from 1978:4 implies a price elasticity of consumption equal to -0.09 after twelve months and -0.19 after twenty-four months. While perhaps a bit on the high side, these are not out of line with other work on this subject.

2. This chapter uses actual stock levels in 1976 as representative of "normal" operating levels in subsequent periods, adjusting proportionally for changes in the level of consumption in the following quarter and also for changes in legal requirements. The choice of 1976 was dictated by *faute de mieux* and by the observation that this was a year of overall supply security and price stability in which no abnormal stock developments were reported.

3. "Deliveries" refers to deliveries of refined products to consumers from refineries. Deliveries are a surrogate measure of end-use consumption, differing from consumption by the amount of change in consumer stocks.

4. Between 1979:1 and 1980:1, the average price of internationally traded crude oil increased at the rate of $1.25 per barrel per month. The cost of holding stocks, including interest and storage charges, was no more than $0.50 per barrel per month during most of this period.

5. Alternative explanations for the cutback are also considered plausible. In particular, it has been suggested that it reflected Saudi displeasure with the Camp David accords, which were signed in March 1980.

3 THE DOMESTIC REFINING INDUSTRY
Economics and Regulation

Edward N. Krapels, E. William Colglazier,
Barbara Kates–Garnick, and Robert J. Weiner

Except for one interlude in 1972–73, from 1931 to 1981 either state or federal governments helped U.S. petroleum refiners obtain access to crude oil. The nature of the assistance varied over the years and at times included additional subsidies for small refiners, but the effect was to allow every refiner to purchase crude oil at a price similar to that paid by other refiners and in quantities to prevent individual capacity factors from falling far below the industry average. From 1931 to 1972, the most significant efforts were made by governments of oil-producing states—preeminently Texas, through prorationing. From 1974 to 1981, the federal regulatory programs were the most important.

As a result, the ability to find and produce crude oil was not a vital prerequisite for entry into the refining industry. Well over 150 refining companies, many of them having no crude oil reserves, were in business in the 1970s.

In 1981, a new era began. The Reagan administration eliminated the federal access programs. State government entities, such as the Texas Railroad Commission, are also no longer involved. Except for the Windfall Profits Tax, the domestic oil market for both crude and products approximates a free market. Refiners with little or no access to their own crude or to long-standing relationships with domestic and foreign crude producers can no longer count on government-guaranteed access to supplies at industry average prices.

55

This change occurred in the middle of a so-called oil glut; that is, a period of excess capacity in the producing countries, reduced demand and large inventories in consuming countries, and falling or weak nominal prices. If there is a future tightening of the oil market from a supply disruption, some domestic refiners will experience reduced crude availability. Which companies are affected will depend on the specifics of the interruption and the steps individual firms have taken to reduce their vulnerability. Without government involvement, the companies with supplies directly or indirectly cut off will have to resort to drawing down inventories, reducing refinery runs, or seeking additional supplies on the spot market. The firms having to resort to the open market are likely to face much higher crude acquisition costs than their competitors.

What are the implications of this state of affairs for U.S. oil consumers? For the refining industry? For the national security in light of the possibility of additional disruptions of foreign crude supplies in the 1980s?

This chapter presents the findings of a study of the "crude oil access problem" for domestic U.S. refiners during a supply disruption (Harvard Energy Security Program 1983). In light of the difficulty of obtaining definitive data describing who gets oil from whom, and on what terms, the study was aimed at identifying the right questions for contingency-planning purposes and at providing answers that are as authoritative as existing data and projections about the oil market would allow. By necessity, our views on many of these questions are accompanied by caveats.

The crude oil access problem is of concern because some regions of the United States might be disadvantaged during a disruption. It is widely perceived that if an independent refiner serving a region cannot get adequate crude oil supplies to produce needed products, then that region might experience considerably higher prices and lower supplies than the national average. States such as Kansas and Indiana, where independent refiners have the largest market shares, might be considered especially vulnerable. In this chapter we do not address the question of whether other firms would be willing or able to move *products* quickly into isolated markets experiencing greater than average shortfalls. We do, however, attempt to place some qualitative bounds on the magnitude of the potential product access problem arising from the crude access problem.

As a working hypothesis we propose the following formulation of the access problem for refiners. In the absence of federal allocation programs, the magnitude of the problem during a disruption is a function of:

1. The extent to which refiners are willing and able to price their products at levels reflecting the marginal cost of crude or the marginal price of finished products;

2. The willingness of producers (for example, Saudi Arabia and Mexico) with incremental supplies to sell to crude-poor firms at or near the industry's average acquisition cost; and

3. The willingness of the U.S. government to sell oil from the Strategic Petroleum Reserve (SPR) to crude-poor firms at or near the average industry acquisition cost or in sufficient volumes at competitive auction to prevent spot prices from rising far above contract prices.

GOVERNMENT CRUDE ACCESS ASSISTANCE IN 1979

The disruption in oil exports from Iran provides a useful case study of the problem and of past government attempts to solve it by regulation. Many American oil companies were directly affected by the loss of Iranian oil. They saw their problem in terms of both prices and quantities due to the existence of price controls in the domestic market. Many refiners wanted crude oil at a regulated price, which was, in the context of domestic controls, the only price that was "competitive" in the sense of allowing the crude-poor firms to create products that could be sold in the American market on a break even basis. Federal regulations helped these firms obtain supplies at such regulated prices.

The Iranian disruption quickly spawned a series of reactions that contributed to an industrywide perception of supply insecurity. The companies that had been direct purchasers of Iranian crude issued *force majeure* notices on their affiliates and third parties with whom they had sales contracts. Some OPEC members, "citing 'extraordinary circumstances', refused to allow many oil companies to lift the full amount of crude oil that was contractually committed to them. By claiming *force majeure*, these countries were able to reap substantial profits from the significant volumes of oil that were then diverted to the spot market" (U.S. Department of Justice 1980:58).

Companies with supply obligations had several ways to allocate the loss. One was to invoke a 100 percent *force majeure* cut to all purchasers. A variant entailed abrogation of third-party contracts while continuing to supply affiliates downstream. A second involved reducing sales volumes on crude from the Middle East, leaving other flows unchanged. This option served to favor importing countries with some domestic production. A third approach was to share the shortage on a pro rata basis among all customers, independently of the type of crude oil and its source (including domestic production). British Petroleum (BP), the largest seller of crude to third parties, set its first priority in satisfying its own affiliates. Just by the cancellation of its long-standing contract with Exxon, it reduced its third-party sales by 24 percent. Exxon's priority appears to have been to supply affiliates while reducing sales to Japan's trading companies. The company allocated oil among its affiliates according to a modification of the International Energy Agency (IEA) sharing formula.

The federal government entered the picture through the Mandatory Crude Oil Allocation Program, commonly referred to as the Buy/Sell Program. This program was mandated under the Emergency Petroleum Allocation Act (EPAA) of 1973 and went into effect in 1974. The objective of the regulations was to equalize capacity utilization across all refiners. This took place through mandatory sales (at average acquisition cost plus handling fees) from "crude-long" to "crude-short" refiners. The regulations were later modified to designate fifteen large companies as sellers, with the price set at the average refiners' acquisition cost of *imported* oil.

In the mid-1970s the world market was calm; that is, the average and marginal costs of imported oil were similar. The Entitlements Program redistributed rents arising from differences between controlled domestic crude oil prices and world prices. In the absence of any further scope for regulation, the Buy/Sell Program languished. Interest in the Buy/Sell Program quickly increased, however, in response to the supply disruption and price instability of late 1978 and 1979. The total volume allocated under the program during the period from the fourth quarter of 1978 (1978:4) to its cancellation in 1981:1 was roughly 215 million barrels. The reason for this increased activity is readily apparent in the divergence between various crude oil costs.

The difference between the average acquisition cost of domestic crude and imported crude increased from $4.48 per barrel in January 1979 to $10.07 per barrel in March 1980. Among the foreign crudes, the price differentials were also significant. The price spread between Iranian and Saudi Arabian oil increased from $0.50 per barrel in January 1979 to $13.65 in December 1979, then narrowed to $5.68 per barrel in March 1980. Importers of Nigerian and Libyan crude experienced an even larger difference vis-à-vis Saudi crude. Tables 3-1 and 3-2 provide data on the wealth transfers under the Buy/Sell Program, estimated to total $1.5 billion. The subsidy to those eligible to buy under the program was calculated by taking the difference between the spot price and average import price and adding $1.50 per barrel as an estimate of the difference between f.o.b. and c.i.f. import costs. The figures should be regarded as rough estimates.

In the absence of any crude allocation, entitlements, or price control programs, companies that could supply their refiners with their own domestic production would obviously have registered greater profits than completely import-dependent firms. Their domestic production affiliates could have priced their crude at or near marginal world prices. Companies with access to lower cost (e.g., Saudi) imports would have had an advantage over companies dependent on high-cost imports.

The Entitlements Program prevented domestic crude producers from realizing the profits they would have obtained in a free market. In that sense, the Entitlements Program was a government-ordained subsidy—paid for by producers—that benefited refiners and consumers. According to an analysis by Kalt (1981), refiners collected roughly 60 percent and consumers 40 percent of this subsidy in 1978.

The only program that had any effect on redistributing the benefits of access to lower cost imported crude was the Buy/Sell Program. In May 1979, the Department of Energy's Economic Regulatory Administration (ERA) spelled out what had previously been an implicit criterion for justifying emergency allocations in 1979—the price of incremental supplies in the world market (*Federal Register* 1979: 26062).

We are recognizing that in cases where a small refiner must pay a price for imported crude oil that is significantly higher than the then current normal range of posted prices in the world markets . . . that small refiner should not be required to purchase this high-priced crude oil. . . . If we did not take into

Table 3-1. Buy/Sell Program—Emergency Allocations.

Month	Million Barrels	Thousand Barrels per Day	Refiners' Acquisition Cost of Imported Oil ($/barrel)	Mideast Light Spot Price ($/barrel)	Subsidy ($/barrel)	Subsidy ($ million)
1/79	1.6	50	15.50	15.95	1.95	3.1
2/79	0.0	0	15.90	19.50	5.10	0.0
3/79	4.5	145	16.40	20.80	6.90	31.1
4/79	2.8	95	17.60	21.20	4.10	11.5
5/79	4.0	130	19.00	34.25	16.75	67.0
6/79	4.3	145	21.05	32.85	13.30	57.2
7/79	5.6	180	23.10	32.00	12.40	69.4
8/79	7.1	230	24.00	32.25	9.75	69.2
9/79	6.2	210	25.05	34.50	10.95	67.9
10/79	8.4	270	25.05	36.00	12.45	104.6
11/79	9.6	320	27.00	39.50	14.00	134.4
12/79a	13.4	435	28.90	39.00	11.60	155.4
1979 Total	67.5	185			11.40	770.8
1/80a	14.8	480	30.75	38.00	8.75	129.5
2/80a	17.1	610	32.40	36.00	5.10	87.2
3/80	13.5	435	33.40	35.75	3.85	52.0
4/80	6.8	225	33.55	35.00	2.95	20.1
5/80	0.2	5	34.35	35.60	2.75	0.5

6/80	1.1	35	34.50	36.00	3.00	3.3
7/80	1.1	35	34.50	33.35	.35	0.3
8/80	1.1	35	34.45	32.30	-.65	—
9/80	1.0	35	34.45	32.25	-.70	—
10/80	1.1	35	34.65	36.80	3.65	4.0
11/80	1.0	35	35.10	39.75	6.15	6.2
12/80	4.6	145	35.65	39.35	5.20	23.9
1980 Total	63.4	175			5.15	327.0
1/81	7.3	235	38.85	39.25	1.90	13.9
2/81	3.2	115	39.00	37.10	-.60	—
1981 Total	10.5	175			1.30	13.9
Grand Total	141.8	180			7.85	1,111.7

a. Figures include allocations to Ashland and Union of 4.8 million barrels (155 thousand barrels per day) in 12/79, 4.7 million barrels (150 thousand barrels per day) in 1/80, and 5.8 million barrels (205 thousand barrels per day) in 2/80.

Sources: Buy/Sell Program, *Federal Register*, (various issues); Refiner's Acquisition Cost, *Monthly Energy Review*, (various issues); Spot Price, *Petroleum Intelligence Weekly*, (1981: 8, 1982: 8). Reprinted with permission.

Table 3-2. Buy/Sell Program—Regular Allocations.

Quarter	Million Barrels	Thousand Barrels per Day	Refiners' Acquisition Cost of Imported Oil ($/barrel)	Mideast Light Spot Price ($/barrel)	Subsidy ($/barrel)	Subsidy ($ million)
1978:4, 1979:1	6.8	37	15.35	16.15	2.30	15.6
1979:2, 1979:3	11.6	63	21.65	31.20	11.05	128.2
1979:4, 1980:1	14.1	77	29.60	37.40	9.30	131.1
1980:2, 1980:3	18.0	99	34.30	34.10	1.30	23.4
1980:4, 1981:1	22.5	123	36.65	38.45	3.30	73.6
Total	72.8	80			5.15	374.8

Note: Prices are averages of monthly data.
Sources: See Table 3-1.

account the price which a small refiner must pay for available supplies, there would be a strong probability that small refiners would encounter strong financial barriers to maintaining the type of viable competitive business operation which the EPAA seeks to protect.

The record of the refining companies that appealed to the federal government for help in 1979 via the crude Buy/Sell and other allocation programs is given in Table 3–3. It provides one measure (that of the ERA) of vulnerability within the domestic refining industry caused by differential crude oil access. It should be remembered, of course, that the U.S. refining industry has not remained static since 1979. Falling demand and decontrol are causing a transition with inefficient or excess refineries being shut down or mothballed at a rapid pace. When such a "sea change" occurs, great care must be taken in drawing conclusions from past vulnerabilities and behavior.

These data indicate that the average refining capacity of firms receiving Buy/Sell assistance was 28 thousand barrels per day, whereas the average size of small refiners receiving no assistance was 15 thousand barrels per day. This suggests that it is too simple to characterize demands for crude allocation programs as coming from the smallest refiners. In fact, in the period from October 1978 to March 1980, the leading recipient of allocated crude was the Texas City refinery, whose capacity is 120 thousand barrels per day. The top ten recipients of emergency allocations—via the Buy/Sell Program only— included Energy Cooperative, Inc., which shut down shortly after decontrol (refining capacity of 126 thousand barrels per day); CRA Inc. (82 thousand barrels per day); Clark Oil (126 thousand barrels per day); Seaview (98 thousand barrels per day); NCRA (54 thousand barrels per day); Rock Island (43 thousand barrels per day); United Refining (42 thousand barrels per day); and Farmer's Union (42 thousand barrels per day).

The three special cases in Table 3–3 merit particular attention. Ashland and Union Oil are large refiners that successfully appealed for special relief from the Department of Energy (DOE). Ashland, which claimed in 1979 to be the largest supplier of independent marketers in the United States, sold off most of its domestic reserves and has obtained the majority of its supplies from foreign sources. In requesting allocations in November 1979, Ashland contended that its refineries were primarily in markets in which it represented over 70 percent of the total refining capacity and 30 percent of oil products sold and that the company was experiencing crude oil contract

Table 3-3.   Refiner Participation in the Buy/Sell Program.

Industry Group	Number of Firms	Refining Capacity (1000 barrels per day)	Share of Total Capacity	10/78 - 3/80 Amount Allocated (1000 barrels)	(1000 barrels per day)	Allocation as Fraction of Refining Capacity
1. Refiners getting buy/sell relief	61	1,764	9.5%	132,158	242	13.7%
2. Special cases[a]	3	1,120	6.1%	20,914	38	3.4%
3. Small refiners not getting buy/sell relief[b]	89	1,362	7.4%	0	0	0
4. Large refiners not affected by allocations	17	2,135	11.6%	0	0	0
5. Refiner-sellers	15	12,008	65.3%	Sales Data: 153,072	280	2.3%

a.   Ashland, Union Oil of California, CORCO.
b.   Less than 50 thousand barrels per day capacity.

reductions and terminations. Ashland maintained that the economic and regional impact of the Iranian disruption would be staggering to its customers. It stated that product pipelines in its refining areas were operating at or near full capacity. Because a hard freeze hindered water transportation, products supplied from Ashland's refineries were said to be especially critical to meet fuel needs of densely populated areas in the Ohio River Valley and other more isolated regions.

Ashland testified that it contacted all fifteen of the largest refiners in the United States about additional crude oil supplies and none offered crude "at any price." Crude oil was available on the spot market, but Ashland said that the price was simply "too high": the price disparity could neither be absorbed by Ashland nor passed on to Ashland's customers. Ashland felt that it could not transfer the burden to its customers because in 1979, Ashland product prices already ranged up to 16 cents per gallon higher than those of the ARAMCO partners and most other refiners.

The DOE Office of Hearings and Appeals granted Ashland 80,000 barrels per day for the period December 1, 1979 to February 29, 1980. DOE's decision caused Ashland to obtain about 9 million barrels of crude oil from selected suppliers. Although the purchase price was the subject of a long subsequent dispute, there is no question that Ashland obtained crude oil at prices well below those prevailing in the spot market.

Commonwealth Refining (CORCO) was not a small refiner; its claim for attention was made interesting by the fact that it was the only refinery in Puerto Rico and was at the time trying to reorganize under bankruptcy laws. In all three cases, the Buy/Sell oil was allocated in a three to six month period, during which it constituted a more significant portion of the firms' total inputs than indicated by the percentages of refining capacity given in Table 3-3.

The fifteen refiner-sellers had to pay the cost of the program. The supplies that were allocated, however, were only a small percentage of their capacity during the October 1978 to March 1980 period. Nevertheless, the cost was very large, as indicated in Table 3-4.

What effect did the program have on oil consumers and on the U.S. economy in general? Little careful analysis has been done on the effect of crude oil allocation on consumers. One study (Verleger 1979) asserts that the allocation program worsened the U.S. gasoline shortages of 1979 because it allocated crude to smaller refiners

Table 3-4. Wealth Transfers from Refiner-Sellers under the
Buy/Sell Program.

Refiner	Subsidy Paid ($ million)
Shell	170
Texaco	170
Amoco	155
Chevron	150
Mobil	140
Gulf	135
Exxon	130
Atlantic Richfield	115
Sunoco	80
Union	70[a]
Phillips	60
Cities Service	35
Marathon	35
Getty	30
Conoco	5
Total	1,485

a. Union also received allocations under the program. These subsidies were worth about $55 million.

Note: Total is from Tables 3-1 and 3-2. Individual company subsidies were calculated by multiplying allocation shares in the *Federal Register* by the total. All figures are rounded to the nearest $5 million.

whose gasoline yield was proportionately lower than that of the major refiner-sellers. According to this analysis, the transfer of 23 million barrels in 1979:2 reduced potential gasoline supplies by 50 to 100 thousand barrels per day.

Another critic of the Buy/Sell Program, William C. Lane, Jr., stated in a report for the American Petroleum Institute (Lane 1981) that the allocation effort "allocated crude away from the East and West coasts, where supplies were short, and deposited it in the Midwest, where supplies were plentiful." In addition, Lane claimed the program "established the precedent of bailing out refiners that had come to depend on the spot market for a large proportion of their crude supplies, reducing whatever incentive they may otherwise have had to stockpile oil for such contingencies in the future."

The benefits of the program claimed by proponents are that the allocation allowed refiners that do contribute to the efficiency of the oil industry in normal periods to weather the disruption and to remain viable enterprises, that allocation of crude oil may have prevented worse regional shortages, and that allocation programs protected the major oil companies from more severe regulatory and legislative changes.

Assuming some government intervention in 1979, an interesting question is whether the United States would have been better off with immediate product price decontrol and a limited temporary allocation of the raw material to a comparatively small number of refiners than with the full regulatory regime of product allocation and price controls. The latter have been almost universally criticized as exacerbating the problems in 1979. Of course, under such a scenario there can be no guarantee that a refiner allocated crude under a limited Buy/Sell Program would supply his old customers, but some of the problems aggravated by excessive government regulations in 1979 would perhaps have been lessened.

THE ECONOMICS OF CRUDE ACCESS

In an ideal competitive oil market, a supply disruption will cause demand initially to exceed supply at the predisruption price. Let us assume that some exporting countries (and owners of crude oil generally) increase their prices faster and further than others, as occurred in 1979. In principle, companies lifting the lower priced crude will be able to collect short-term rents, and their internal transfer prices will be set so as to reflect the greater scarcity of supply in the market. The refiners without access to the lower priced supplies would go to the open (arm's-length or spot) market, where they would pay prices that reflect oil's increased scarcity. *These rents have no effect on economic efficiency, the market price, or competition in the industry.*

In principle, the companies with access to lower priced oil should be indifferent between selling crude on the spot market (to oil-poor refiners) and passing the oil downstream within their own corporate networks at market-equivalent transfer prices. Hence, the oil-poor refiner should be able to remain competitive with refiners with access to lower priced oil. All firms would base oil prices on the cost of the marginal barrel. The prices of feedstock going into the refineries would be equivalent, taking into account quality differences. The only difference is that some refiners collect higher rents on the crude than others. The consumer is charged the market clearing price.

There are several reasons why the real world may not conform to this model. The reasons can be divided into three groups:

1. Government imposed price regulations or fears thereof;
2. The tendency of large firms to allocate resources internally by administrative fiat rather than according to short-term profit-maximizing considerations; and
3. Predation (the lowering of price below marginal cost in order to encourage exit from the industry).

The last of these has received the most publicity. Critics of the major oil companies have often charged that in the absence of government intervention, the better-established integrated companies would use a period of tight supplies to squeeze the independent refiners out of business or into a style of management that conforms more closely to their desires. This view is based on the belief that under normal market conditions it is the independents that drive product prices down to competitive levels. To drive out or discipline these competitors, the argument goes, the majors will price products below the independents' marginal costs in a disruption.

For an integrated company to seek to squeeze an independent out of the market through predatory conduct, anticipated future gains, discounted appropriately for risk, must exceed the profits forgone in the predatory period. For these profits to exist, there must exist elements of market power in refining—otherwise attempts to raise prices will be unsuccessful—and there must be barriers to entry in the refining industry. It would hardly be worthwhile to squeeze out an independent today if another (or the same one) can easily be in business tomorrow.

The relationship between competitive behavior in the very short and very long runs is in general indeterminate. Inasmuch as a supply disruption is characterized by high oil company profits and great uncertainty about the future, it appears a particularly unpropitious time for employing a strategy of trading present sacrifice for anticipated future gains. On the other hand, entry into the refining business is not costless, and established integrated firms could benefit in the postcrisis period if a number of their independent competitors had abandoned the field during the disruption.

Another, less dramatic, view of why firms with access to lower priced crude might not behave in accordance with the model stresses the role of government intervention and of companies' perceptions thereof. According to this view, even without formal price controls, an oil disruption will encourage the government to exert influence on the private sector via jawboning and the threat of reregulation. The petroleum industry is always an inviting target, and the larger enterprises have often been singled out for attack. It is noteworthy that even in such an avowedly free-market country as West Germany, the government put pressure on oil companies in 1979 to "moderate" their price increases. At the end of the year, Chancellor Schmidt responded to reports of profit increases by accusing the companies of making fools of Europeans, stating, "I regard these exorbitant profits as provocative."

The third reason that companies with access to lower priced oil might price their supplies below marginal cost deals with the propensities of large companies to allocate supplies within their own organizations and to historic customers by administrative fiat rather than short-term market considerations. In the 1973–74 crisis, and again in 1979, many integrated international firms carried out a pro rata supply reduction to industrialized countries. The market would have produced such an allocation only if these countries had identical demand elasticities, which is unlikely. Multinationals' allocation efforts affect each of the national markets involved. Part and parcel of this crisis management philosophy is a refusal of firms to sell to those outside the group of historical customers. Moreover, because of the constraints of existing contracts and commercial practice, companies—especially those not subjected to *force majeure* cancellations of supply—might not be able to break contracts and sell to new customers at higher prices, even if they wished to do so.

History suggests that oil companies with access to lower priced crude will not necessarily price their products at marginal costs. One does not have to embrace theories of anticompetitive behavior to subscribe to this view, and the lower prices would benefit consumers. *There is no escaping the fact, however, that such behavior by firms with access to lower priced crude creates problems for firms which must purchase on the arm's-length markets.*

Economics suggests that companies will raise prices quickly to market-clearing levels. Historical behavior in the oil industry indicates otherwise, but history may not be an accurate guide, since the current domestic free market is far different from the comprehensive regulatory regime existing in 1979. There is still concern among some companies that, because of contract rigidities, the market mechanism will not take care of all the short-run problems; that is, companies with crude might not be able to move products around quickly to severely impacted regions.

THE OIL MARKET IN THE 1980S AND THE THREAT OF SUPPLY DISRUPTIONS

A BP executive in early 1982 said that the world was experiencing its first demand crisis. World demand has fallen dramatically, in part due to conservation and structural changes in energy use in response to high oil prices, but also to the worldwide recession and stock withdrawals. To date, the data do not exist to separate out definitively the various influences of price, policy, inventory drawdowns, and recession. The long-term price elasticity of demand has proved to be much higher than previously believed.

The pressure of falling oil demand has been placed on OPEC with the other oil producers operating near maximum levels. The existence of excess production capacity is clearly a significant difference from the 1970s. The temporary existence of large stocks and surge capacity, as well as reduced imports from OPEC, means that the world is temporarily less vulnerable to a supply interruption. It could be argued that the world would not be much affected by a disruption in any country except Saudi Arabia.

Even without any political calamities, many economists expect that the oil market will tighten sometime in the mid- to late 1980s because demand is projected to increase in the oil-producing and oil-importing less developed countries. Thus, the expectation is that the world will experience a slack period for several years, with flat or falling real prices, until output begins to approach production capacity.

The domestic market is changing as well. Fifty years of state and federal regulation of the crude oil market is partly responsible for the proliferation of nonintegrated refiners in the United States. Falling demand and free market forces are reshaping the U.S. oil industry.

Most companies are struggling to make refining and marketing operations profitable in a shrinking market. Many refiners and marketers have announced their intention to pull out of both certain geographic areas and certain unprofitable classes of trade. American Petroleum Refiners Association data indicate that between 50 and 60 of the U.S.'s 324 refineries have shut down since January 1981. According to a Phillips Petroleum executive: "The ones that survive can and will be profitable. But we are not there yet. We have not reached the new equilibrium point."

To investigate the current state of affairs regarding crude access for domestic refiners, one would ideally survey who buys crude from whom and the terms of such arrangements. Nevertheless, we can infer that domestic crude is disposed of in one of the following ways:

1. Production by affiliates of integrated oil companies. Most of the large integrated refiners produce oil. A few produce more oil than they refine. Most have domestic self-sufficiency ratios of less than one, and many of them are looking to increase the security and diversity of their supplies—as with Gulf's attempted purchase of Cities Service.
2. "Locked in" purchases by refiners from independent crude producers. Some independent producers sell only to selected refiners as a result of geographic and physical connections or as a result of having long-term fixed price contracts—probably a rarity these days.
3. Purchases from independent producers under "evergreen" contracts with periodic (quarterly or, more typically, monthly) price negotiations.
4. Spot purchases.

We suspect that the volume of crude moving through the third channel has increased at the expense of the second. During the 1970s, domestic producers of price-controlled crude had no incentive—and in some cases were barred by the supplier/purchaser freeze—to switch buyers. With decontrol, the producers have an incentive to enter into more flexible contracts.

It is unlikely that differences in access to the crude of domestic independent producers will cause an access problem in the 1980s. Evidence suggests that independent domestic producers will price their crude at world market levels in a tight market. Both majors and independents will face these market prices. Problems may occur if majors with substantial domestic crude production subsidize either the acquisition of incremental supplies by not basing their intrafirm transfer decisions on marginal costs or subsidize product prices by not basing them fully on marginal costs.

The international market is the source of the incremental barrel for domestic refiners. *As such, the behavior of foreign producers is the single most important determinant of the crude access problem for American refiners in a supply disruption.* The fact that U.S. imports are now a lower percentage of refinery runs, however, means that the problem is of less concern than in the late 1970s.

Ironically, the problem would be mitigated if all foreign producers behaved in accordance with the classical model. To the consuming countries' advantage, however, countries like Saudi Arabia effectively controlled the price of their crude in the 1979 disruption. Since 1979, the international crude access picture has become more complicated than it was in the days when each country had a set group of concessionaires. Today, traditional crude oil contracts still constitute the most important channel of supplies, but the contracts are generally for much shorter periods. In some cases, old relationships have been able to withstand market gyrations (e.g., the ARAMCO partners and Saudi Arabia); others have been found to be less stable (e.g., Gulf's relationship with Kuwait).

A disruption will cause the greatest difficulties for companies with two types of relationships in the international market: first, companies unlucky enough to be customers of the country experiencing the disruption; and second, companies dependent on spot market purchases. During a disruption, a two-tier price structure, with a difference of $10 to $20 per barrel between contract and spot prices, as occurred in the past, is a likely prospect. It should be remembered, though, that firms buying on the spot market in early 1982 gained a financial advantage over those forced to pay higher contract prices. And there is no guarantee that establishing term contracts with supposedly reliable foreign producers is the most prudent course. If Saudi Arabian production were lost—the most serious event for the current oil market—the integrated ARAMCO partners would be seriously affected, and the remaining OPEC producers might price all their supplies near the spot level.

IMPLICATIONS AND REMEDIES OF A CRUDE
OIL ACCESS PROBLEM IN THE 1980S

Let us assume that those companies maintaining access to lower cost crude supplies may, for a variety of reasons, refrain from pricing their products at levels reflecting the marginal cost of crude on the world market or the marginal price of finished products. The reasons might include: informal government pressure, avoiding the image of price gouging during a national emergency, being forced to pass savings through by a price-moderate foreign producer, or attempting to gain a larger market share for later advantage. The customers of those companies would reap the temporary benefits. Historical evidence would tend to support the likelihood of this "moderate" price behavior; but, as we have emphasized, generalizing from a period of price controls may be misleading.

Independent refiners seeking incremental crude supplies will likely have to pay spot prices. Those competing in regional product markets with integrated refiners that charge less than the marginal cost for products would be forced to absorb short-term losses, curtail throughput, or go out of business altogether. The customers in this area would benefit in the short term from the lower prices. In principle, the exit of these independents from the refining industry could affect the competitive nature of the market in the future.

Those independent refiners normally serving a major fraction of a regional product market will be able to stay in business by significantly raising prices. Thus, these regions may suffer higher prices and lower supplies than other parts of the country. However, because crude can be obtained on the spot market, these price differences would probably be no higher than the difference between the spot price and average industry price. At $10 to $20 per barrel, this amounts to $0.25 to $0.50 per gallon of gasoline.

Product price differentials much above transportation costs, however, would not persist if marketers outside the affected region move products in quickly in an effort to make a profit. Whether this will happen in the short run we cannot say, but in the long run it will certainly happen. The constraints of existing contracts and commercial practice may be the most significant obstacle in preventing products from moving quickly to severely impacted regions. But the oil industry has a vested interest in ensuring that political pressures do not force the government to step in with a regulatory regime that may outlast the crisis. For that reason, oil companies may go to great lengths to ensure that no one "goes cold" and that all sectors and regions have a chance at obtaining what supplies are available.

Thus, the essence of the crude oil access problem is that some regions of the country may experience in the short run somewhat higher prices and lower supplies than other areas. Congress and the public are very sensitive to perceived inequities during a crisis, which creates pressure for government involvement. Of course, political problems and large economic losses arise from the sharp increase in average product prices, but this is irrelevant to the *differential* price problem. For oil companies, the crude access problem is summarized by the spread between spot and contract crude oil costs and the pricing policies of companies with access to less expensive crude.

Although we have not conducted a formal survey of the programs that other countries have to deal with the access problem, it is our impression that virtually all IEA member countries have standby allocation programs of one kind or another. In some countries, governments oversee intraindustry cooperative efforts that would be in violation of U.S. antitrust laws. In others, there is in addition a mandatory stockpiling requirement that gives companies some breathing room when their supplies are disrupted. And one major, BP, has proposed a private sector initiative whereby companies can purchase contingent claims on each other's inventories (up to five days) in order to prevent bidding on the spot market by the companies most severely affected by the disruption. In other countries, governments in a crisis spearhead the search for incremental supplies by seeking direct purchases from exporting countries.

If the U.S. government perceives crude oil access as a problem, it could respond either by disbursing oil from the SPR or by implementing a limited Buy/Sell Program for refiners. Although the SPR is designed primarily to reduce economic losses by moderating the average oil price increase, the U.S. government could also eliminate the differential crude access problem by selling oil in sufficient volumes at competitive auction to replace much of the U.S. shortfall or by selling oil to the crude-poor firms at prices at or near the industry average. The amount sold in the first option would have to be sufficient to mitigate substantially the upward pressure on world market prices, which is possible only for limited disruptions. To make this option work, the SPR sale—whether it were a "physical" or a "futures" sale—would have to be done promptly and expertly to avoid private inventory accumulation. The second option would be an acknowledged subsidy to the affected refiners. It would discourage those refiners from holding high stock levels and diversifying their sources in normal times, thereby providing the wrong incentive for reducing vulnerability.

The SPR could also be used if the IEA sharing agreement is triggered. A key question facing the federal government is how to structure the interaction between a domestic free market and a possible international allocation system. Compatibility could be ensured by using the strategic reserve to compensate companies when their supplies are directed by the IEA to other countries.

The second alternative, a limited crude oil sharing program, would require reestablishing a standby, limited Buy/Sell Program, which could also be used if the IEA system is triggered. Great care would have to be taken to draw up rules that avoid the mistakes of the allocation programs of the 1970s. It is not certain that such a limited program could remain limited; it may start the country once again down the road to comprehensive price controls and government allocations. Both market (SPR auction) and nonmarket (SPR allocation or limited Buy/Sell Program) remedies are available if the effects of the crude oil access problem are perceived to be serious enough for government intervention. The costs and benefits of both types of remedies need to be addressed in greater detail.

REFERENCES

Federal Register 44, no. 88 (May 4, 1979): 26060–66.

Federal Register. 1978–81 (various issues).

Harvard Energy Security Program. 1983. "Crude Oil Access in Disruptions in the 1980s: Analysis of Public Policy Implications." Harvard Energy Security Program Discussion Paper Series, H-83-03.

Kalt, Joseph P. 1981. *The Economics and Politics of Oil Price Regulation: Federal Policy in the Post-Embargo Era.* Cambridge, Mass.: MIT Press.

Lane, William C. 1981. *The Mandatory Petroleum Price and Allocation Regulations: A History and Analysis.* American Petroleum Institute. Mimeo, May.

Petroleum Intelligence Weekly 20, no. 5 (February 2, 1981): Special Supplement.

_____. 21, no. 10 (March 8, 1982): Special Supplement.

U.S. Department of Energy. 1974–81. *Monthly Energy Review* (various issues).

76   THE OIL MARKET

U.S. Department of Justice. 1980. "Report of the Department of Justice to the President Concerning the Gasoline Shortage of 1979." Washington, D.C.: U.S. Government Printing Office, July.

Verleger, Philip K. 1979. "The U.S. Petroleum Crisis of 1979." *Brookings Papers on Economic Activity*, No. 2: 463–76.

4 THE ROLE OF INTERNATIONAL COOPERATION

Ulf Lantzke

International cooperation on oil supplies is, of course, not new. The Organization for European Economic Cooperation (OEEC), the predecessor of the Organization for Economic Cooperation and Development (OECD), set up the Oil Committee in 1948. The committee played a major part in the international coordination of the reaction to the supply crisis following the closure of the Suez Canal in 1956 and the resultant imposition of rationing in Europe. Suez was a transport crisis with the closure of the canal resulting in the need to send tankers around the much longer Cape of Good Hope route. It was solved in a relatively short time by increased U.S. output from shut-in production which was then exported to Europe. Following this experience the procedures that had been developed for sharing oil supplies between European countries were formalized and supposedly available for use in the next crisis. Agreement was also reached in 1958 on desirable stockholding levels for the European members of the OECD (then the OEEC).

However, in 1973 it was not so easy. Following the outbreak of the Yom Kippur War in October 1973 the Arab oil producers systematically reduced production and embargoed supplies to the United States and the Netherlands. The United Kingdom and France, by contrast, were supposed to be given special treatment. The result was international paralysis, and the OECD Oil Committee was unable to agree even to the formal collection of the information necessary to operate an oil sharing scheme. Instead, there was an intense scramble for oil, prices rocketed, and the companies had the highly distasteful task of trying to allocate their inadequate supplies between their clamoring customers in a reasonably fair manner.

The result of this experience was the initiative of the U.S. Secretary of State, Henry Kissinger, which led to the drawing up of the International Energy Program (IEP) and the creation of the International Energy Agency (IEA) in November 1974. The IEA binds its signatories[1] to cooperate in an oil-sharing scheme that would come into effect in the event of a severe oil supply shortfall either for the member countries as a group or for any individual member or members. To assist in withstanding any such shortfall each participating country commits itself to holding emergency reserves—basically oil stocks—sufficient to sustain consumption for at least ninety (originally sixty) days with no net oil imports.[2] Some of the key features of the sharing scheme that I would like to mention are that:

1. It cannot be activated unless the shortfall in total oil supplies to the IEA group of countries or an individual country is *expected* to be more than 7 percent. This means that it is not sufficient to look at a supply interruption in one or more producing countries and say that the sharing scheme must be triggered. An assessment must also be made of the likelihood of supplementary production elsewhere. In practice, therefore, the 7 percent hurdle is a very stiff one to pass. Thus, in January 1979 the loss of some 5 million barrels per day (mbd) of Iranian exports was equivalent to some 10 percent of world oil supplies but, because the shortfall was partially compensated by increased production from Saudi Arabia and elsewhere, there was never any question of the IEA as a group of countries suffering a shortfall of more than 7 percent. In current circumstances, with substantial spare capacity in most OPEC countries and, to a much more limited extent, in non-OPEC countries such as Canada and Mexico, even a total disruption of Saudi production would not necessarily trigger the scheme.

2. If the executive director makes a finding that the threshold has in fact been reached, activation, though not a formality, is almost certain to take place, because of the voting arrangements built into the IEP.

3. In that event, member countries have a further obligation to impose demand restraint measures sufficient to take care of the first

7 percent or, if the crisis is big enough, the first 10 percent of the supply shortfall.

4. Beyond the demand restraint commitment the remaining short-fall is shared out between member countries through the re-allocation of supplies according to a formula. This imposes a substantially greater stock drawdown on countries that are totally dependent on imports than on countries that are partially or fully self-sufficient.

5. The actual process of re-allocation requires the close cooperation of the oil industry, which would send a substantial team to Paris to advise the secretariat.

Although the sharing scheme has not as yet had to be implemented it has been regularly tested and will be tested again next year. Despite numerous criticisms of the scheme from different quarters, the twenty-one member governments of the IEA and the forty-nine oil companies who would be directly involved are firmly committed to it and we are strongly of the belief that, if called upon, it would work.

IEA member governments have undertaken these commitments in the expectation that the benefits of cooperative action—both political and economic—would far outweigh any costs. The impetus for creation of the IEA was in fact more political than economic. The 1973–74 Arab oil embargo brought home to the oil-importing countries the political vulnerability associated with import dependence. The IEP was conceived with the aims of reducing *vulnerability to dependence* in the short run and reducing *dependence itself* in the long run.

Although the threat of another deliberate, targeted embargo by oil exporters appears more remote today than it did in 1974, the political benefits of the IEP are still important to member countries. During a major world oil supply crisis, intense political pressure would be placed on all governments to assure their citizenry and industries adequate supplies of oil at a "reasonable" price. While most leaders would recognize that competitive bidding against other importing countries would leave all importing countries worse off, this logic might not so easily prevail in the domestic debate. In the absence of international agreements and commitments to eschew such competition, competition is likely to be the path of least resistance for many politicians. The resulting strains on relations among competing countries could likely be severe, and these strains in turn are likely to interfere with the importing nations' ability to cooperate in other areas at what would surely be a time of extraordinary international tension.

It is also worth pointing out that the oil industry, whose active participation would be vitally necessary to successful implementation of the IEP, also believes it would derive important benefits from participation.[3] During the 1973–74 Arab oil embargo, with no IEP in effect, the companies found themselves in a difficult situation. Their basic objective was to share available supplies equally among their local affiliates. At the same time, officials of governments in whose jurisdiction the companies were based were quick to summon company officials to discuss the supply situation both globally and in the local context. At these meetings, the company men would have much preferred to be in a position to explain to governments that, notwithstanding their own unfailing patriotism, the allocation among countries of the oil under their control was dictated strictly by the arithmetic of the IEP, backed up by the force of international law, and therefore could not be altered.

Turning to the economic benefits, the events of 1979 underscored the economic losses that can be sustained as a consequence of disrupted oil markets. As a result, member countries are probably more mindful today of the potential economic benefits of the IEP than of the political benefits. These benefits would result from the restraining effect of the Emergency Sharing System on world oil prices both during and after a disruption. The drawdown of emergency reserves would augment available supplies and reduce the shortfall. Where market forces are relied upon domestically to allocate supply shortfalls, the drawdown of emergency reserves would enable the market to clear at a lower price. In countries where rationing or other forms of nonprice allocation are favored, the drawdown of emergency reserves would reduce the percentage reductions that must be applied to various classes of consumption.

On the international oil market, the demand restraint, reserve drawdown, and sharing provisions of the IEP would in combination reduce upward pressure on prices. The first two components would reduce aggregate IEA demand for imports, while the third would nullify any incentives for one member to increase its share of the group's imports by bidding against another member of the group. (The incentive to bid against nonmembers would remain but would be diluted somewhat since success against nonmembers would have to be shared equally with other members.)

The immediate benefits of lower prices on the world market would be the smaller transfers of income to producers during the disruption. These benefits would continue well beyond the actual disruption period since, as has now become apparent, it takes years for market forces to begin to undo the sharp price increases that can result from a few months of supply uncertainty. Furthermore, since rigidities in our economies cause oil price increases to result in lower levels oi GNP, another important benefit of restraining oil price increases would be to reduce the GNP losses that we would otherwise see.

IEA RESPONSE TO "SUBTRIGGER" DISRUPTIONS

Neither of the two oil supply disruptions that have occurred since the IEP was established in 1974 has been large enough to trigger the Emergency Sharing System.[4] Instead, IEA member governments responded collectively and individually with a variety of ad hoc measures. During the first of the two subtrigger disruptions, these included a few instances in which national self-interest was placed before that of the group as a whole. Such instances were notably absent during the second disruption.

The most important IEA action of 1979 was a March 2 declaration that:

> IEA countries will contribute to a stabilization of the world situation by reducing their demand for oil on the world market. The reduction would be in the order of 2 mbd, which would correspond to about 5 percent of IEA consumption. Each participating country will regard this as a guideline in the policies it will pursue to achieve its contribution to this reduction. These policies are expected to yield equivalent results in participating countries.

Measures to achieve this reduction in demand on the world market were envisaged to include fuel switching, short-term conservation (building temperature controls, etc.), increased indigenous production, and drawing on stocks.

While some governments, including the United States, did take domestic actions aimed at reducing consumption in the months following the March IEA meeting, these were neither as stringent nor as urgent as would have been necessary to achieve the 5 percent objective through nonprice means. Instead, the average price of crude oil on the world market rose from $14 per barrel in the first quarter of 1979 (1979:1) to $24 per barrel in the fourth quarter of 1979 (1979:4). Consumption in 1979:4 was 2 percent less than in 1979:1 on a seasonally adjusted basis, and 4 to 5 percent less than it would have been had there been no disruption and price increase. The objective was therefore achieved but, unfortunately, the cure was the very disease that governments had sought to avoid.

While the primary cause of the runup in prices was the scramble by oil companies to assure supplies, a few government actions exacerbated the situation. Several governments initiated direct state-to-state contacts with crude oil producers to acquire supplies for state-owned refining companies, and the U.S. government subsidized the import of heating oil by $5 per barrel. While the effect of these actions on the market was probably not significant, they nonetheless demonstrated the weakness of the IEA commitment to collective action. A second initiative in the form of import ceilings for 1980 was agreed upon at the Tokyo summit meeting in June 1979, but this turned out to be largely an empty gesture. The ceilings were aimed not at the current market but at the following year and were set at comfortably high levels to avoid the difficulties of defining equality of sacrifice for countries with such diverse energy requirements.

In a sense, the positive return on the efforts of 1979 came later, in September 1980, when governments were suddenly faced with a second disruption of equal magnitude. Having had time to reflect on the dynamics of the oil market during the previous disruption and on the inadequacy of their own actions to control events, member governments were now determined to take a more decisive posture. Within seven days of the outbreak of hostilities between Iran and Iraq, the IEA Governing Board agreed that member governments would adopt the following measures:

- Urging and guiding all public and private market participants to refrain from abnormal purchases on the spot market;

- Immediate consultations by Member Countries with oil companies to carry out the policy that, in 1980:4, there would be a group stock draw sufficient to balance supply and demand, taking into account whatever additional production is available to the group;

- Consultations between member governments of the IEA to ensure consistent and fair implementation of [the above] measures taking into account market structures in individual countries, and to adjust for imbalances which might occur in particular situations;

- Reinforcement of conservation and fuel substitution measures which are already contributing to lower demand for oil.

These measures were reaffirmed by the Governing Board meeting at the ministerial level on December 9, 1980, and their meaning was amplified in two respects. The first measure, counseling against "abnormal purchases" on the spot market, was restated as "discouraging undesirable purchases at price levels which have the effect of increasing market pressure, with a view to removing as much buying pressure as possible on a broad basis"; and the reference to adjusting imbalances (third measure) was expanded by adopting a measure to "contribute to the correction of supply imbalances, through a system of consultations between governments, companies, and the [IEA] Secretariat."

A comparison of the emphasis of the March 1979 and October 1980 Governing Board decisions shows something of the lessons learned in 1979. The emphasis in March 1979 on reducing demand for imports had reflected the belief that the problem was merely one of reducing demand (consumption) to equal supply (production). By contrast, in October 1980, while the last of the measures pertained to restraining consumption, the emphasis had clearly shifted to an attempt to control the *demand for oil in excess of amounts needed for consumption, namely, the demand for stocks.* The post-mortem on 1979 had shown that production and consumption had not been seriously out of balance at any point during the year but that steady stockpile accumulation had accounted for the buying pressure after 1979:1. The objective of the IEA Governing Board in October 1980 was to ensure that stocks would this time serve to reduce, rather than exacerbate, the market imbalance.

The October 1980 measures also reflected a recognition that, even while the global supply–demand balance might be satisfactory, localized imbalances could nonetheless cause market pressure. It was felt that if countries were being asked not to search for new sources of supply during the crisis, those among them with serious supply difficulties would have to be helped out in some other way. The commitment to sharing among member countries was thus seen as an important part of the overall plan.

It turned out that Turkey, which had been purchasing over 50 percent of its supplies from Iran and Iraq, was the only member country whose supply situation was urgently affected by the war. Turkey did not ask for formal activation of the IEP's selective trigger provisions (although the gravity of its supply situation gave it a right to do so under the terms of the IEP) on the understanding that oil would be provided by less formal means. Efforts on the part of the IEA secretariat and officials of the United States and the United Kingdom governments to find companies willing to sell oil to Turkey were hampered by several factors: reliable information on Turkey's exact requirements proved difficult to obtain as the supply pipeline from Iraq was operating on and off during November, December, and January; companies were concerned about Turkey's ability to pay; and the year-end holidays intervened at the most critical moment, making it difficult to coordinate negotiations between Washington, London, Paris, and Ankara. In early January, when sellers had been located and price terms were being discussed, the pipeline reopened and the supply crisis receded without any offers being accepted.

The fact that the world oil market remained far calmer in the months following the Iran–Iraq War than in the months following the Iranian Revolution had at least as much to do with the difference in oil market conditions as with the difference in government responses. In the first case stocks were low and demand was rising; in the second case stocks were high and demand was falling. These differences made it much easier for companies and governments to honor the spirit of the IEA's October 1980 decision. Nevertheless, that decision was taken, and governments were clearly prepared to do everything possible to prevent its being flouted.

THE DEVELOPMENT OF "SUBTRIGGER" POLICIES IN THE IEA

As early as the spring of 1980, well before the second subtrigger disruption, work had begun within the IEA secretariat to put proposals to the Governing Board for a more specific system for responding to disruptions below the 7 percent threshold for emergency sharing. A preliminary step was taken by the Governing Board in May 1980, when a system of periodic consultations between governments and companies on inventory policies was agreed upon. But the next step—a political consensus on a concrete program of action in the subtrigger situation—proved elusive, and the differences of views among member countries widened further in January 1981.

The lines of this debate have often been characterized as a simple opposition between laissez faire and *dirigiste* schools of thought. But while these philosophical leanings have played their part in the discussion, especially in its rhetoric, a number of practical, and therefore somewhat complex, proposals were put forward and equally practical objections raised.

The IEA secretariat's proposals, as discussed at several meetings of member government officials and with oil industry representatives, can be summarized as follows:

- Continuous monitoring of the oil market by the secretariat, together with establishment of an information system to operate during nondisrupted periods, designed to provide the Governing Board with a rapid, accurate appraisal of the situation should a disruption occur;

- At the onset of a disruption, rapid consultation by the Governing Board to consider adoption of appropriate measures. Such measures would include:

  -discouragement of abnormal spot market or other undesirable purchases;

  -short-term demand restraint and fuel switching;

  -increased indigenous production;

  -the use of stocks;

  -efforts to minimize and contain the effects of supply imbalances.

In implementing any of these measures, governments would consult with oil companies in their jurisdiction.

The first point was broadly supported and has in fact been put into effect. The secretariat now receives regular monthly reports on oil supplies to member countries in the current month, the previous month, and the first future month. Using this and other sources, the secretariat publishes a monthly oil market report.

On the second point, the only consensus that could be achieved among member governments was that the Governing Board should meet promptly to provide a forum for sharing information. Beyond this, however, there was a strong hesitation on the part of more than one delegation to endorse any measures in advance, even if these were understood to be merely a list of possibilities to be considered in light of actual circumstances.

The Use of Stocks

In the discussion of individual measures, by far the greatest interest was directed toward the use of stocks, reflecting the experience of 1979 and the recognition of their crucial role during that period.

Before detailing various possible approaches to stocks policy it is well to be aware of the differing stockholding arrangements in IEA countries, since this to some extent determines government attitudes. In the European Economic Community (EEC) countries—nine members of the IEA and France—the legal obligation is ninety days of last year's consumption, with an offset of fifteen days in the case of the United Kingdom to allow for North Sea production. Japan has a similar ninety-day requirement. However, in North America the IEA requirement of ninety days of last year's net imports is below the industry's minimum operating requirements, and there is no other obligation as in Europe and Japan. Instead, there is in the U.S. case the Strategic Petroleum Reserve (SPR), now standing at 310 million barrels or twenty-one days of forward consumption. There are also much smaller volumes of strategic stocks in Germany and Japan that are nevertheless of the same order of magnitude as in the United States in days of consumption. Then there are the national oil companies such as Petro Canada, ENI, and Petrogal, over the stocks of which governments obviously have some control. In Europe and Japan, however, the legal obligation to hold ninety days of consumption is specifically geared to an oil crisis involving the IEA sharing scheme, while in the United States the administration has also made it clear that the SPR is reserved for such circumstances. Similar concepts are involved with the strategic stocks in Europe and Japan. Thus, there is at present no block of government-controlled oil stocks readily available for use in tight market situations falling short of the 7 percent threshold.

Industry or Government?

In the autumn of 1980 governments had some, although not a wholly unqualified, success in persuading companies to draw down stocks to meet the supply disruption rather than to purchase on the spot market or through premium-priced contracts. But this was partly because in a series of consultations earlier that summer the companies had made it clear that they were becoming worried that inventories were too high in relation to demand and wished to run them down. Consequently, when the disruption came, there was a certain commonality of interest between the companies and the governments. In the further round of consultations with companies in 1981, however, it became clear that as a general principle they did not relish being ordered about in possibly noncommercial ways in relation to what is nowadays a significant part of their total asset base. Company cooperation in the use of commercial stocks to dampen tight market situations could therefore not generally be relied upon. Nevertheless, the EEC compulsory stockholding requirements are a partial exception. The companies have indicated that they would welcome the additional flexibility of being able to use, say, five days out of their EEC obligations in a tight market situation to meet their own requirements. They would not, however, be so willing to make this stock available to other companies or governments, and so the value of such flexibility could be limited.

Consequently, while not ruling out governmental consultations with companies designed to encourage them to use stocks in particular ways in particular situations, most recent analysis has tended to focus on the use of government stocks, assuming that these could be made available.

Apart from the problem of cost, the most important difficulty raised against proposals to use government stocks concerned decision-making: How, in the early stages of a supply disruption, when the future course of the crisis is extremely uncertain, can governments prudently decide to use the stocks they own or control? And if the situation is such that governments believe they *can* prudently do so, why wouldn't the private market participants share this perception and use their own stocks, making government intervention unnecessary?

Some respond that the current ninety-day emergency reserves are adequate for a true emergency, so that any additional stocks governments might choose to create for use in the preemergency period could be used without risk. Others argue that the potential economic damage from a subtrigger disruption is sufficiently large in itself to justify use of some portion of the emergency reserves—say five of the ninety days—in the preemergency period.

In my own view, these arguments rest on a false formulation of the question. Whether it is prudent to draw down stocks is a relevant question for an individual company or country concerned with its own supply situation, since oil drawn down by one entity is no longer available to that entity. But from the global standpoint, the drawdown of stocks by any individual entity, whether a company or a country, does not in itself reduce the total amount of stocks available to the group. If governments sell their stocks or release mandatory stocks, they are merely transferring ownership and control to the companies. Since the volume of stocks remains the same, the overall level of protection remains the same. What has changed is the companies' propensity to look for additional protection, and this is precisely the desired effect.

Over time, if the release of government stocks is successful, prices will be lower than they otherwise would have been, consumption will be higher and, if production remains unchanged, stocks will be lower. Only in this indirect and delayed fashion would the release of government stocks reduce the total amount of stocks. But within the short time frame of a supply disruption, if the concern is that a serious situation might deteriorate into an emergency, release of government stocks in the preemergency period would in no way diminish the overall level of protection available for coping with the emergency.

Volumes and Costs

The volume of oil required for the effective use of stocks to dampen the oil market is significant, but the cost is *not* significant if spread over the totality of oil use. Twenty-five to thirty million tons (six to seven days of IEA consumption) would have been sufficient to cover the physical shortfall during the 1979 and 1980 supply disruptions. A willingness to use a larger stock of, say, 50 million tons or twelve days of consumption would also have had a powerful psychological effect. But, although the capital cost of purchasing and storing 25 million tons of crude could be 5 to 6 billion dollars, the annual financing and running costs would amount to an almost imperceptible one cent per U.S. gallon of IEA gasoline consumption, or 0.3 cents per U.S. gallon if spread over all products. The question, therefore, should not be whether a security tax of this kind can be afforded, but whether the stocks concerned could really be used effectively.

Ways of Moving Stocks Into the Supply System

If government stocks are available and the decision to move stocks into the supply system is made, there are, in principle, a number of ways in which this could be done:

1. Governments sell stocks to companies operating in their domestic market to offset those companies' loss of imports: This is the most straightforward use of government stocks and is akin to the anticipated use of the SPR to meet severe supply disruptions. Governments would either physically transfer stocks to the companies for their own use or permit them to count in some part of government stocks for the purposes of meeting their mandatory stockholding obligations.

2. Governments deal directly with other governments: One government could arrange to sell or loan oil to the government of another country in need, or to an oil company for delivery to that country. The objective would be to make up a supply deficit that might otherwise have to be covered from the spot market.

3. Governments release stocks to domestic suppliers who are willing to direct imports to other countries in need: This method would take advantage of the ability of many international companies to transfer oil quickly from one destination to another without substantially disrupting their supply networks. It would be particularly appropriate in cases where the company operated in both of the countries concerned.

4. Governments sell oil directly on the spot markets: In the event of governments holding a relatively large volume of stocks, the knowledge that they were willing and able to release oil into the spot

market should have a highly beneficial effect on the psychology of the market and discourage the speculative element that is present in any supply disruption.

5. Governments make oil available to a central body for disposition: I mention this to complete the picture and because it could in theory give the greatest flexibility of all. However, in our world of nation states, delegation to a supranational body is unlikely to be acceptable.

Pricing

One difficult problem on stock transfers is pricing. The ideal in terms of limiting price increases would be to price at predisruption official or spot prices. However, this becomes difficult once the market price has moved and might, in any event, lead some operators to buy the oil but simply hold it in the hope of obtaining a substantial stock profit rather than deploying it where needed. One practical approach might be to charge at or just below the day's market f.o.b. price with freight and other allowable costs specified. This should encourage buyers by reflecting current market conditions without adding to upward price pressures. Where governments are releasing oil to domestic suppliers the auction method may be thought desirable. The difficulty is that, although stock transfers should ease price pressures by working on the underlying causes, high auction prices could give the contrary impression. "Lending" oil against an undertaking to redeliver like physical qualities in the future might be one way of avoiding the pricing issue.

International Coordination

All these methods of moving stocks into the supply system, and pricing, would need some degree of international coordination to make the initial judgment as to when to move oil into the market, monitor the results, and agree on further corrective action if necessary. The appropriate forums for such coordination are the IEA and the EEC. On the other hand, it might not be necessary to involve all the member countries of the IEA and the EEC. In consumption terms the United States, Japan, Germany, France, Italy, Canada, and the

United Kingdom account for 80 percent of the OECD consumption and, if stockholding were proportionate to consumption, would in any event more or less carry the burden.

Trading In Claims on Emergency Reserves

I would like now to return to the major exception to my original proposition that the use of government stocks is a more promising avenue than company stocks in a tight market situation: Some months ago, officials of an international oil company put forward a proposal for an oil contingency scheme run entirely on commercial lines. The main outlines of the scheme are that:

- An independent agency would be established, with borrowing powers on the financial markets. (There would be some governmental representation on the board.) The agency would purchase up to five days worth of oil at market-related prices from the participating companies, which in turn would acquire corresponding claims.

- In the event of a participating company being able to show that its normal contracted or equity supplies have been disrupted by *force majeure* it would be able to buy back its claims at the previous month's average purchase price. It would, however, undertake to resupply the agency, at the same price, within six months.

- It would also be allowable for companies to trade in claims, thus providing a further adjustment process both prior to and during the disruption. Trading prior to the disruption would allow companies to invest in or divert claims according to their perception of the risk inherent in their respective supply positions. Trading during a disruption would allow affected companies to buy claims from unaffected companies, which would be ineligible to exercise the claims. In this way, more oil would be mobilized than if governments simply allowed all companies to use five days of emergency reserves at their discretion.

- The running costs of the scheme would be met entirely by the participants, with no costs falling on governments. However, in order to give an initial impetus to get the scheme off the ground, it is suggested that in Europe governments should agree that the inventories held in the pool should count against a corresponding proportion of importers' existing compulsory stock obligations, thus mobilizing stocks that are at present frozen. In the United States and Japan it might in principle be possible to formulate similar but tailor-made schemes with, in the United States, the possibility of the oil being held as part of the SPR.

This proposal should have much the same effect as would the use of government stocks in that it should discourage a surge of buying by the affected companies on the world's spot markets in the event of a short-term disruption. However, it also has the further powerful advantages of requiring neither government money nor government decisions to operate successfully. Indeed it would be essential that governments should not interfere in its day-to-day operations. It is an industry-born proposal designed to make the oil market more "complete" in a technical sense and the operation of market forces more efficient.

I do not know whether this proposal will come to fruition. It may be that the economic incentives are insufficient for the companies to be willing to take it up of their own accord. However, since oil supply disruptions affect not just the oil market but the whole world economy, governments would do well, both individually and in the EEC and the IEA, to look very carefully at the ideas behind the scheme and consider whether they could do anything to give it a helping hand.

CONCLUSIONS

The problems of managing oil supply disruptions have for the moment been moved offstage as the public debate focuses on the possibility of a collapse of oil prices. Yet, even with the current volume of spare production capacity, the loss of production and production capacity in the Gulf would interrupt 13 percent of free world oil supply requirements for 1983 after all spare capacity outside the Gulf was brought into service. Taken together with the continuing fragility of the political situation in that region, this fact demonstrates the continuing importance of the IEP to the economic and therefore political security of the oil-importing nations.

I would conclude with a plea that the respite provided by the current slack oil market be put to good effect. Of course, the next crisis may never come. But the postwar history of the Middle East, combined with the present spate of cancellations of projects that would substitute for OPEC oil and the blunting of the conservation drive as the real oil price falls, suggests that it very well may. In my view we are quite well prepared for the really major crisis for which the IEA sharing scheme is designed (although there is always scope for further improvements). But we are still *not* well prepared for lesser disturbances. It would be a tragedy if the events of 1979 and the economic recession that has ensued had to be repeated once again for the lessons to be fully learned.

NOTES TO CHAPTER 4

1. Australia, Austria, Belgium, Canada, Denmark, Germany, Greece, Ireland, Italy, Japan, Luxembourg, the Netherlands, New Zealand, Norway, Portugal, Spain, Sweden, Switzerland, Turkey, the United Kingdom, and the United States.

2. In December 1982, the Governing Board agreed that member countries would make efforts to maintain stocks equal to ninety days of average imports over the previous three years, except where consumption had declined because of clearly established long-term structural change.

3. The forty-nine major oil companies who participate in the IEP do so on a voluntary basis.

4. The shortfall during 1979:1, when the Iranian Revolution reduced exports to zero, was about 2 mbd, or 4 percent of free world consumption. In 1980:4, when the war between Iran and Iraq interrupted production in both those countries, the shortfall was again in the vicinity of 4 percent.

OIL SUPPLY SHOCKS AND MACROECONOMIC POLICY

5 ENERGY SHOCKS AND THE MACROECONOMY

Robert S. Pindyck and Julio J. Rotemberg

Future energy prices are highly uncertain. The rate of discovery of new energy supplies and the development of new energy technologies are highly variable. The continued ability of OPEC to agree on production allocations and avoid competitive price cutting is questionable. Added to this is the inherent instability of the Persian Gulf and the world oil market in general. As a result, the "confidence interval" around any forecast of future energy prices must be a large one.[1] In the future energy prices may rise or fall unpredictably, and as we have seen in the past, this could have important implications for the performance of the American economy and the other industrial economies.

This chapter has three objectives. The first is to provide a brief overview of the various ways in which changes in energy prices—gradual changes or, alternatively, sharp and sudden "shocks"—affect macroeconomic variables such as inflation, employment, real output, and investment. The second objective is to summarize the results of our recent work on the dynamics of energy use in industrial production and to discuss what those results show about the effects of energy price shocks on investment, employment, and industrial

Work leading to this paper was supported by the Center for Energy Policy Research of the MIT Energy Laboratory, and that support is gratefully acknowledged. Our thanks to Stan Fischer and Bill Hogan for helpful comments.

energy use. The last objective is to discuss the implications for both economic and energy policy.

We should stress at the outset that when we discuss energy shocks, we are referring to price shocks as opposed to unexpected shortages of energy. Barring a major war with widespread disruption of oil shipments, shortages are almost impossible at the level of the world oil market. As long as some oil is being produced and shipped, any country can import as much as it demands by offering a high enough price. The function of the world oil market, and in particular the spot market, is to let prices equalize supply and demand.[2] Shortages at the retail level, on the other hand, are a distinct possibility and can result from government attempts to hold fuel prices below market-clearing levels. Fortunately, the United States has now removed price controls on crude oil, and interstate natural gas markets will be in large part deregulated by 1985. Unless controls are revived during periods of rapid increases in energy prices, a repetition of the gasoline and natural gas shortages the United States witnessed in the past is unlikely. But although shortages are unlikely (or rather, would be of our own making), price shocks can occur quite easily; they are the focus of this chapter.

An energy price increase can have two different effects on an industrial economy, and it is useful to distinguish between them. First, a higher price of energy has a direct effect by reducing the total real national income available for domestic consumption and investment. It does not matter whether the cost of energy rises because it is imported and a cartel raises its monopoly price or because depletion of potential and proved reserves makes domestic energy sources more difficult to tap. Nor does it matter whether the price increase occurs slowly or rapidly. In each case the higher cost of energy will mean a lower real national income—that is, a loss of real purchasing power. This in turn will mean lower real wages, profits, and consumption levels. Furthermore, no economic policy can eliminate this direct effect. The reduction in real income will occur even if monetary and fiscal policies are used to keep the economy close to full capacity output and employment.

The second effect of energy price increases—the adjustment effect—occurs when those increases are rapid and unexpected. An energy price shock will raise the rates of inflation and unemployment, and reduce investment levels. This will cause a further change in real national income and may well magnify the direct reduction discussed above. The adjustment effect occurs because of the rigidities that characterize our economy—rigidities in prices, in the use of inputs to production, and in wages. Adjustment effects cannot be eliminated entirely, but they can be significantly reduced through the proper use of economic policy. The wrong policies, however, can magnify adjustment problems such that they become a serious threat to economic growth and stability.

Below, we take up the direct effects of energy price increases. We then turn to a discussion of adjustment effects. Finally, we examine the implications for economic and energy policy. We will argue that the proper economic policy response to an energy shock is continued moderate growth of the money supply, restraint on fiscal expansion, and, if possible, reductions in payroll and excise taxes. As for energy policy, we argue that a tariff on imported oil should be imposed now, before another shock occurs. Such a tariff could be reduced during a shock, thereby insulating the economy from sharp price changes. A large tariff may be preferable to a major expansion of the Strategic Petroleum Reserve. Also, we see no need for "recycling" schemes that channel money back to consumers in the wake of a price shock; such schemes would be counterproductive.

DIRECT EFFECTS OF RISING ENERGY PRICES

The direct effect of an energy price increase is simply an implication of the fact that more domestic resources must be "traded" for each Btu of energy. The magnitude of this direct effect therefore depends on the structure of energy demand—in particular, on the cost share of energy in GNP and the ability of household and industrial consumers to substitute away from energy when it becomes more expensive.

The cost share of energy as a fraction of GNP sets an upper bound on the extent to which an increase in the price of energy will reduce real national income. If there were no substitution possibilities (i.e. energy demand were completely price inelastic), and if the supplies of capital and labor were constant, output would remain fixed, and the drop in real national income would be $S/(1-S)$ times the percentage price increase, where S is the share of energy in GNP (about 8 percent in the United States). In other words, for a 1 percent increase in energy prices, an additional $S/(1-S)$ percent of our (fixed) output would have to be traded for the same amount of energy. The impact on real national income would be smaller the more elastic the demand for energy. A very rough estimate of the overall price elasticity of energy demand in the United States is about -0.6, which would imply that a 10 percent increase in the price of energy would reduce real national income by about 0.6 percent.[3]

The analysis above does not depend on whether energy is used as an intermediate input or as a final product. In particular, even if all energy is used by households for heating and transportation, if the amount of energy consumed does not change in response to price changes, the fall in real national income is still given by $S/(1-S)$ times the percentage price increase.

It should be kept in mind that while the direct effect of an energy price increase implies a reduction in real national income, it need not imply very much of a reduction in real GNP or, therefore, in productivity. This is particularly the case if much of the energy is imported. The reason is that GNP is a measure of final output and not the real flow of goods and services available for domestic use.[4] If the elasticity of substitution between energy and other inputs to production is small, domestic production will be largely unaffected by higher energy prices. The problem is that more of that production must be bartered away for the more expensive energy; this is why real *income* falls. It is interesting to note that the larger the elasticity of substitution between energy and other production inputs, the larger will be the reduction in real *GNP* resulting from an energy price increase, but the smaller will be the reduction in real income and, thus, our standard of living.

For this reason it is important to focus on the real income losses connected with energy price increases, and not on GNP or productivity losses. For example, policies that speed up the replacement of energy-inefficient capital (e.g., a gas-guzzling car) with energy-efficient capital can reduce productivity because they accelerate the use of energy-efficient but labor-inefficient technologies. But for an oil-importing country, this means a reduction in imports and improvement in the terms of trade, so that output may fall, but income will rise.[5] In the long run, the problem with steady increases in energy prices is not that they will drain away our output and productivity growth but that they will drain away our real incomes and living standards.

Besides having a direct effect on national income, energy price shocks also influence the demand for energy itself and the demands for various types of durable goods. Moreover, the initial impact of a price shock may be followed by various repercussions. Consider, for example, the reaction of households to higher energy prices.

Households use energy mainly for transportation and heating. Their demand for energy is thus a demand for services that depend on durable goods, such as houses, cars, and refrigerators. As is well known, different durable goods of the same class (e.g., cars) yield the same services when combined with different amounts of energy. Thus, an unexpected increase in the price of energy has two effects. First, it reduces the use of the existing stock of durable goods, thus reducing energy consumption. Second, it changes the demands for different types of durables; that is, it raises the demand for goods such as insulation and energy-efficient cars. However, it will take time for the composition of the stock of durables to change in response to this change in demand, so in the short run, the relative prices of durables will be affected. For example, owners of large automobiles incur capital losses while owners of energy-efficient appliances accrue capital gains.

The capital gains and losses would be smaller if people knew ahead of time when the energy price increase would occur. If people knew now that the price of energy will increase at some point in the future, they would alter their purchases of durable goods now and this would induce a change in their demand for energy. This suggests that one should not strive to obtain a single price elasticity of the demand for energy. The response of energy demand depends not only on the short run versus the long run. It also depends on people's perception of future energy prices and, in particular, on inferences about future energy prices based on observations of current ones.

These dynamic effects of energy price shocks are just as important when the industrial sector is considered. In general, a changing energy price will imply a change in the mix of productive factors that firms employ. It may, for example, lead to less investment and reduce capital accumulation in the future. For this reason it is important to understand the ways in which an energy price shock will affect the demands for factor inputs to production.

ENERGY PRICE CHANGES AND INDUSTRIAL FACTOR USE

Estimating the direct impact of energy price changes on investment behavior, employment, and energy use requires a description of the

structure of industrial production, namely, a model of factor demands. If the energy price change is sharp and sudden—that is, if it
takes the form of a price shock—it is essential that such a model be
dynamic and that it account for anticipations regarding future prices.

To see why, note that in response to sharp energy price changes
firms want to change not only their energy purchases but also their
use of capital. Insofar as it is costly to change the stock of capital
quickly, capital responds slowly to such sharp price changes, and the
long-run elasticity of energy demand is different form the short-run
elasticity. Much of the disagreement over recent estimates of industrial energy demand elasticities has focused on whether these estimates pertain to the short or long run. Moreover, the size of the
costs of adjusting capital also influences the extent to which energy
demand—and the demands for other factors—respond to prospective
changes in the price of energy.

In a recent paper (Pindyck and Rotemberg 1983) we developed a
dynamic factor demand model that retains the generality of func-
tional form that has characterized much of the recent static modeling
work; we used a translog restricted cost function. In addition, that
model is consistent with producers holding rational expectations and
optimizing dynamically in the presence of adjustment costs. We estimated that model using aggregate data for the U.S. manufacturing
sector, treating energy and materials as flexible inputs, but capital
and labor as quasi-fixed—although we found adjustment costs on
labor to be very small.

That model can be used to show how factor inputs change over
time in response to various kinds of anticipated and unanticipated
changes in the price of energy as well as changes in the price of capital, the level of output, and so on. Here we summarize the results of
using the model to simulate the effects of the following "events":
(1) the price of energy unexpectedly increases by 10 percent and is
then expected to remain at this higher level; (2) the same 10 percent
increase in the price of energy is anticipated by firms five years
before it occurs; (3) the price of energy rises gradually at 2 percent
per year; (4) the cost of capital unexpectedly declines by 5 percent.

The results of simulating these four hypothetical events with the
model are shown graphically in Figures 5-1 through 5-4. Figures
5-1a through 5-4a show percentage changes in capital, labor, energy,
and materials inputs over time. These percentage changes are relative to the base year 1971, at which time all factor inputs were taken

to be in steady-state equilibrium. Figures 5–1b through 5–4b show the behavior over time of the ratio of net investment to the capital stock. For simplicity, inputs and output are assumed constant in steady-state equilibrium, so that in the absence of an event, net investment is zero.

Figures 5–1a and 5–1b show the effects of an unanticipated 10 percent increase in the price of energy. The major impact is a significant drop in the use of both capital and energy, which in the model are complements. Because of adjustment costs, capital falls gradually, while energy, a flexible factor, falls by a significant amount in the first period and continues to fall in subsequent periods in conjunction with the drop in the use of capital. The parameter estimates of our model indicate that adjustment is fairly rapid; about three-fourths of the total drop in capital occurs in seven years.

Suppose that same energy price increase is anticipated five years before it occurs. As can be seen in Figures 5–2a and 5–2b, the same steady-state equilibrium eventually results, but the dynamics of adjustment are quite different. The demand for capital begins dropping immediately because of the presence of adjustment costs, but the major changes in the use of energy and labor occur after the price increase is realized. There is still a noticeable decline in the use of energy prior to the price increase, however, because of the complementarity of energy and capital. Note that simulations such as this one can be used to analyze the impact of the 1978 Natural Gas Policy Act and related legislative future changes in energy prices.

In Figures 5–3a and 5–3b, the price of energy increases steadily at 2 percent per year. Costs of adjustment are then negligible, and, as one would expect, the result is gradual changes in the values of all factor inputs.

One can also use a model such as this to examine the effects of changes in the prices of other factors. Figures 5–4a and 5–4b show the simulated effect of an unanticipated 5 percent decline in the cost of capital (resulting, say, from a change in the tax laws). The result is a substantial increase in the use of capital (in the model, the long-run own price elasticity of capital is nearly −3.0), but also a significant increase in the use of energy, which in the model is complementary to capital.

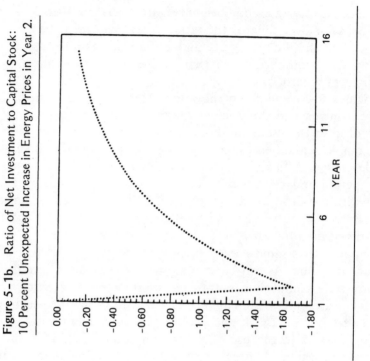

Figure 5–1b. Ratio of Net Investment to Capital Stock: 10 Percent Unexpected Increase in Energy Prices in Year 2.

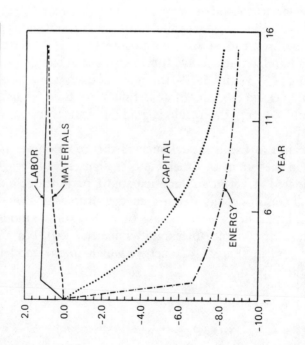

Figure 5–1a. Percentage Changes in Input Use: 10 Percent Unexpected Increase in Energy Prices in Year 2.

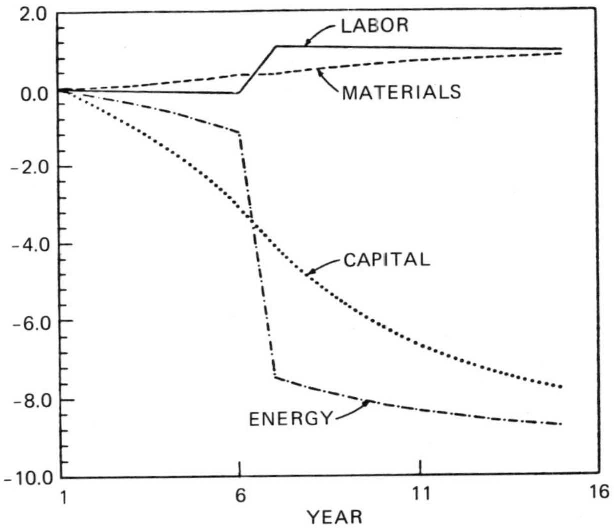

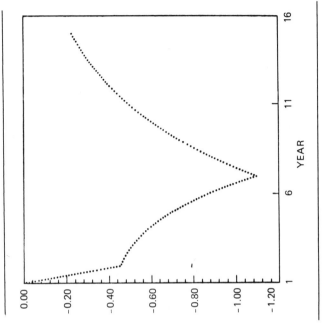

Figure 5–2b. Ratio of Net Investment to Capital Stock: 10 Percent Increase in Energy Prices Beginning in Year 7, Anticipated as of Year 2.

Figure 5–2a. Percentage Changes in Input Use: 10 Percent Increase in Energy Prices Beginning in Year 7, Anticipated as of Year 2.

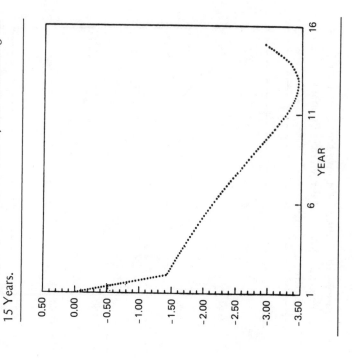

Figure 5–3b. Ratio of Net Investment to Capital Stock: 2 Percent per Year Increase in Energy Prices Beginning in Year 3, Expected as of Year 2, and Continuing for 15 Years.

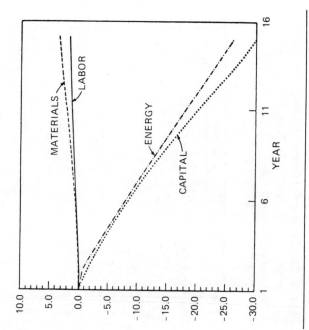

Figure 5–3a. Percentage Changes in Input Use: 2 Percent per Year Increase in Energy Prices Beginning in Year 3, Expected as of Year 2, and Continuing for 15 Years.

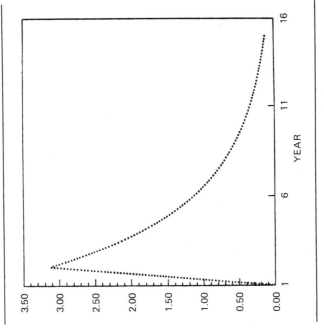

Figure 5-4b. Ratio of Net Investment to Capital Stock: 5 Percent Unexpected Decrease in the Cost of Capital Beginning in Year 2.

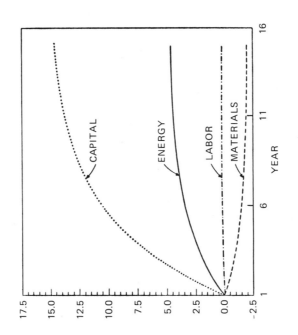

Figure 5-4a. Percentage Changes in Input Use: 5 Percent Unexpected Decrease in the Cost of Capital Beginning in Year 2.

GENERAL EQUILIBRIUM EFFECTS

The model used for these simulations has an important limitation: It is a partial equilibrium model. In other words, when simulating the effect of an increase in the price of energy, the real wage rate and the level of output were taken to be fixed. In fact, these variables will also change, leading to further changes in the use of labor and capital and, thus, in the use of energy.

Even if all other factor prices remain constant, an increase in the real price of energy leads to an increase in the marginal cost of production at an unchanged level of output. This induces firms to reduce their output, thus further reducing their demand for energy and also reducing the demands for the other inputs to production.

While the real prices of capital and materials might be relatively unaffected by an increase in the price of energy—since they too are produced using capital, labor, energy, and materials—the real wage can be influenced in a number of ways. First, the demand for labor is likely to decline for any real wage. It is conceivable that the desire on the part of firms to substitute labor for energy may be so strong that the demand for labor actually rises in spite of the output effect mentioned above. However, this is unlikely because if the supply curve for labor were upward sloping (or vertical) and unaffected by the change in energy prices, the real income of workers as a whole would increase in response to the higher price of energy. This would require a sharp drop in the income of the owners of capital to ensure that national income actually fell. Therefore, it is more plausible that the demand for labor falls. This in turn induces a fall in the real wage, if the labor supply schedule is unchanged by the increase in the price of energy.

The supply of labor itself may fall for any given real wage measured in domestic goods, as explained in Branson and Rotemberg (1980). As the real price of energy rises, the real compensation of a worker who consumes energy directly falls, even if the ratio of his wage to the price of domestic output is changed. This may induce some workers to substitute nonmarket for market activities. The resulting shift in the supply of labor raises real wages but lowers employment and is thus consistent with the fall in total real labor income. It must be noted that this shift in labor supply, while theoretically plausible, has not been isolated empirically, thus lending credence to the notion that real wages must fall in equilibrium. This fall in real wages induces a further fall in the demand for energy as firms substitute energy for labor. But this is somewhat offset by an increase in the demands for all factors as the fall in real wages lowers marginal costs of production. The magnitudes of these effects is a subject of ongoing research.

Finally, an increase in the relative price of energy can affect the level of nonenergy prices. The size and magnitude of this effect depends on the characteristics of the demand for money. As we saw, a direct effect of the increase in energy prices is a fall in real income. However, the demand for real money balances depends on the level of income. As real income falls people demand a smaller quantity of real money balances. To maintain the demand for the existing level of nominal money balances, the price level must rise. The price level, however, is a composite of energy and nonenergy prices. Therefore, nonenergy prices will tend to rise more the more responsive is the demand for money to changes in income, the sharper is the fall in real income, and the lower is the weight of energy prices in the price level.

This does not guarantee that nonenergy prices will rise; in theory they could fall. If the demand for money were completely inelastic with respect to income, the increase in energy prices would have to be offset by a decrease in nonenergy prices in order to keep real money balances constant and equal to the constant money demand. In practice, we expect money demand to be sufficiently elastic so that nonenergy prices will indeed rise, as they did following the 1974 and 1979–80 oil price shocks.

ADJUSTMENT PROBLEMS

So far we have discussed the direct effects of energy price changes. If those price changes are sharp and unexpected, as they were during the past decade, they will also cause significant adjustment effects, at least during the short run. These adjustment effects will come on top of the reductions in real income and output discussed earlier.

Adjustment effects are the result of rigidities in our economy—in particular, rigidities in prices and wages. For example, after an energy price increase, the prices of other goods may not adjust rapidly to their new steady-state values. In addition, real wage rates may fail to fall quickly to the new lower equilibrium level that is likely to be required by higher energy prices. In this case labor prices itself out of the market; from the point of view of firms, full employment becomes uneconomical; and the unemployment rate rises and real output falls.

This problem is particularly acute in Europe where, as Sachs (1979) and Branson and Rotemberg (1980) have shown, real wages tend to be rigid. The problem is that most workers would rather not hear the message that real wages must fall. Since for many European workers wages are automatically linked to prices through explicit and implicit cost-of-living escalators, the message can be ignored. Wage-setting processes based on cost-of-living escalators will, in the wake of an energy price shock, cause wages to rise when instead they may well have to fall relative to the cost of living. This contributes to the inflationary spiral, lowers employment, and increases unemployment as workers price themselves out of the labor market. These changes in employment lead to a reduction in real output. The problem is that the economy is unable to come back to equilibrium quickly at a lower real wage level.

As Sachs (1979) and Branson and Rotemberg (1980) have shown, in the United States it is the nominal wage and not the real wage that is sticky. This is due to the presence of long labor contracts that are signed in nominal terms and have only partial escalator clauses. Since, for the reasons discussed above, there tends to be a burst of inflation following an energy price shock, such a shock leads to a fall in the real wage, as an unchanging nominal wage must now be deflated by a higher price level. Indeed, as Table 5-1 illustrates, the real wage fell significantly in the United States during the 1974 and the 1979-80 shocks.

The question remains whether these declines in real wages were smaller or larger than those that would have prevailed in the absence of sticky nominal wages. This is an important empirical question. If in the short-run real wages in fact fall more than they would in the absence of nominal wage rigidities, then output falls by less.[6] The opposite occurs if nominal wage rigidities prevent real wages from falling as much as they should.[7] These effects follow from the incentives firms have to lay off workers when real wages are high and to hire them when real wages are low.

Rigidities in nonenergy prices create further adjustment effects on output, and the direction of these effects is also ambiguous. As

Table 5-1. Percentage Change with Respect to the Previous Year of
Real Hourly Earnings of Nonagricultural Employees.

Year	Percentage
1972	3.0
1973	-0.1
1974	-2.8
1975	-0.7
1976	1.4
1977	1.0
1978	0.5
1979	-3.1
1980	-4.0

Source: *Economic Report of the President* 1982: 276.

explained above, in the absence of rigidities the nonenergy component of the aggregate price level could go up or down depending on
the characteristics of the demand for money and the extent to which
real income changes. If, in the absence of rigidities, nonenergy prices
would fall, then the presence of rigidities would delay this fall; real
money balances would temporarily be lower, thereby reducing people's purchasing power; and hence, output would temporarily be
lower. The opposite would be the case if nonenergy prices would rise
in the absence of rigidities.[8]

Price rigidities can create a further effect on output. It is possible
that nonenergy firms will raise their prices too much by just "passing through" increases in energy costs without regard to demand conditions. Then, output will be lower in the short run until firms finally
adapt to depressed demand by lowering their prices.

The point here is that rigidities in nominal wages and nonenergy
prices can lead to short-run shifts in real output, beyond the direct
reductions discussed earlier. Ongoing research will hopefully help to
establish the direction and magnitudes of these short-run shifts and
the extent to which adjustment problems exacerbate the economic
effects of energy price increases.

One last effect should be mentioned. Real output can also be reduced in the longer run because of the dampening effect that an
energy price shock can have on investment. We saw earlier that one
of the direct effects of an energy price increase is to alter the mix of

factor inputs in industrial production, reducing investment to the extent that energy and capital are complementary. In addition to this, a sudden change in energy prices can create uncertainty about the profitability of private investment. Combined with the higher interest rates that can accompany the higher rate of inflation, this can bring about a slowdown in investment demand.[9]

IMPLICATIONS FOR ECONOMIC POLICY AND ENERGY POLICY

An energy price shock can result in a combination of higher inflation, higher unemployment, and lower real output, and this complicates the design of an economic policy response. There will be social and political pressures on the one hand to pursue contractionary policies in order to fight inflation, and on the other hand to pursue expansionary policies in order to stimulate output and reduce unemployment.

Following the 1974 oil price shock, the goal of fighting inflation dominated. The Federal Reserve used a highly contractionary monetary policy as a means of "treating" the inflationary burst that followed the oil price shock, and this resulted in a recession that was probably worse than it needed to be.

On the other hand, an expansionary monetary policy is likely to postpone the fall in real income that is necessitated by the increase in energy prices. It must be reiterated that it is still an open empirical question as to whether nominal rigidities in the United States tend to raise or lower output relative to what it would have been otherwise. Monetary policy can, of course, take advantage of these rigidities and temporarily increase output by following expansionary policies. However, since the direction in which output should move in the short run is ambiguous, we would recommend that money growth simply be smooth and predictable so as to provide a stable economic environment and to control inflation in the longer run.

An energy price shock can also create problems for fiscal policy. As we explained earlier, an energy price shock must be followed by an adjustment in real national income, and this means adjustments in the components of real national income, including the government sector. Unfortunately, if real income falls after a shock, there is a tendency for the share of government in the economy to rise. The role of government and the size of government's share in GNP should be determined by social and economic considerations. Any reductions in real national income following a price shock should be spread evenly across all sectors of the economy, including the government sector. Otherwise, government budget deficits can grow out of control, leading to still further inflation in the future as well as a more than proportionate drop in real income levels outside the government sector. Because of this, an energy price shock should not be followed by a cut in income taxes. In fact, if the government wants to keep expenditures in real terms at their preshock levels, tax *increases* will be needed.

Some people have argued that a system should be put in place to "recycle" purchasing power back to consumers in the wake of an energy price shock. We see no need for such a scheme, because the government would have no additional revenue to "recycle" unless it imposed new taxes. An energy shock would increase revenues from the Windfall Profits Tax, but these are likely to be more than offset by the reduction in income tax receipts. Also, the recession that accompanies an energy shock would increase the "uncontrollable" outlays of the government (e.g., unemployment compensation). "Recycling" schemes that have been proposed are simply methods of increasing government transfers. They have no rationale as part of economic policy or energy policy.

Incentives to stimulate investment are desirable following an energy price shock. Policies such as investment tax credits and accelerated depreciation can be used to help stimulate investment demand, which might otherwise be depressed. This can help to accelerate capital-energy substitution earlier and thereby reduce energy consumption, since new capital is likely to be more energy efficient.

Insofar as nonenergy prices rise too much, it is also desirable to utilize cost-reducing tax policies, which lower prices and raise output by reducing costs of production. Such a policy provides a way for the government to compensate for the flexibility that the economy lacks in the short run. The best candidate for this kind of policy is a reduction in payroll taxes. A payroll tax increase has the same kind of effect as an energy price increase—it raises production costs, causing more inflation, which in turn feeds back into wages, and so on. It is unfortunate to note that we are moving in the opposite direction, with major payroll tax increases scheduled.

In summary, while it is impossible to eliminate the inflationary and recessionary impacts of an energy price shock, there are policies that can be used to ameliorate the impact and prevent it from becoming worse than necessary. Such policies include moderate growth of the money supply, restraint on fiscal expansion, investment incentives, and reductions in payroll taxes.

The macroeconomic impact of an energy price shock also has implications for energy policy. Perhaps most important, by now we should have learned the lesson that the imposition of price controls following a shock can be disastrous. Price controls can turn a price shock into a quantity shock—that is, they can cause significant shortages of energy. The macroeconomic impact of higher energy prices can be large but not nearly as large as the economic impact of energy shortages.

The macroeconomic impact of an energy price shock should be viewed as a social cost, since no single firm or consumer can affect that impact. As such, it warrants a public policy response. In particular, our vulnerability to sharp increases in the price of oil adds a premium to the social value of every barrel of oil.

In addition to economic vulnerability, our dependence on imported oil also creates an undesirable political and strategic vulnerability. Because of the economic and political consequences of a major cutoff of OPEC oil, we find ourselves spending vast sums to develop a "Rapid Deployment Force." We are also subject to political pressure by oil-exporting countries. These political and strategic costs further raise the premium on the value of a barrel of oil.

Estimates of the size of the premium vary, depending in part on one's view of the political costs of import dependence, but a reasonable estimate would be in the range of $10 to $20 per barrel. Some people have argued that this premium justifies the subsidization of synthetic fuel development as an eventual substitute for oil. But this is not the case, and such subsidies would be an extremely inefficient way to deal with our import dependence.[10] Rather, the premium implies the need for a tariff on imported oil.

Unlike a "Btu tax" or related excise tax, a tariff would raise the price of all oil in the United States, affecting both the demand and supply sides of the market. It would give consumers an added incentive to conserve, and it would give producers an added incentive to produce—efficiently choosing those energy sources (oil or others) and those technologies that are most economical. At the time of an energy shock, the tariff would be reduced or temporarily removed, thereby insulating the economy from a sharp energy price change.

Unfortunately, in the short term a tariff—like an OPEC-induced increase in the price of oil—is inflationary and recessionary. But this problem can be largely eliminated if the tariff is phased in gradually over a two- or three-year period. (Remember that gradual increases in energy prices are much less damaging than rapid increases.) Also, imposition of the tariff can be combined with reductions in the payroll tax.

The present Windfall Profits Tax would tax away part of any increase in revenues accruing to oil producers as a result of the tariff. Whether that tax should be expanded or reduced as part of a tariff program is a question that we have not yet addressed in detail; it hinges on both equity and resource depletion considerations.

The "social premium" on oil may also justify the strategic oil reserve. A strategic reserve has two functions. First, in the event of a war that disrupted most shipping and trading of oil, shortages could occur in that imports might be unavailable at any price. Strategic reserves could be used to prevent such shortages. Second, a strategic reserve could be used to smooth out the price increases that follow a major production cutback, thereby reducing their economic impact.

But a strategic reserve works best when implemented as part of an international agreement. When a stockpile is released, wherever it happens to reside, it adds to the world supply of oil and thereby reduces the world price. As a result the benefits are enjoyed by all importing countries, even if they do not have stockpiles of their own, but the benefits to the country holding the stockpile are reduced. If only the United States released a stockpile in the wake of a crisis, the impact on world oil prices—and prices faced by Americans—would be small. But if most major Organization for Economic Cooperation and Development (OECD) nations maintained and released large stockpiles, this would significantly limit price increases and the economic damage they cause. Chapter 11 examines this issue in detail.

As the experience of the International Energy Agency has shown, the likelihood of obtaining such an international agreement is low. Meanwhile, the costs of implementing the strategic reserve have risen. The program should therefore be carefully reassessed. It may be better to aim for a smaller stockpile but impose a hefty tariff on imported oil.

Table 5-2. Summary of Energy Price Shock Effects.

Effects	Policy Implications
I. Direct Effects	
1. Reduction in real national income. Size depends on energy cost share and demand elasticities.	Economic policy cannot reduce this effect. However, to keep government share fixed, government spending must fall commensurately.
2. Short-run drop in investment spending. Behavior of factor demands over time depends on pattern of price change and on anticipations.	
3. Real output unchanged if energy demand is price inelastic and home factors are supplied inelastically; otherwise real output falls.	
4. In general, real wage will fall. Theoretically possible for real wage to rise if (a) factor substitution effect is very strong, or (b) large drop in labor supply as workers substitute leisure for labor.	
II. Adjustment Effects (Assumes real wage falls in 4.)	
5. Burst of inflation in short run. Inflationary burst is greater the greater the direct reduction in real income and the more elastic the demand for money with respect to changes in real income.	Inflationary burst is largely unavoidable. Do *not* respond with a sharply contractionary monetary policy.

6. If the real wage is rigid (e.g., Europe), output falls by more in short run, and employment falls. — Reduce payroll or value added taxes. Use tax incentives to stimulate investment.

7. If the nominal wage is rigid (U.S.), inflationary burst reduces real wages.
(a) If real wage falls by more than it would in absence of rigidity, then output falls by less. — Contract money supply.
(b) If real wage falls by less than in absence of rigidity, output falls by more. — Expand money supply.

8. If nonenergy prices are rigid, and they:
(a) rise by less than they would in absence of rigidity, then output falls by less. — Contract money supply.
(b) rise by more than in absence of rigidity, output falls by zero. — Expand money supply.

Since 7(a) vs. 7(b) and 8(a) vs. 8(b) are empirically unresolved issues, money growth should remain stable.

III. Energy Markets

9. Price controls will turn a price shock into a quantity shock, which has much worse economic impact. — Avoid the use of price controls or "allocation" schemes.

10. Macroeconomic adjustment effects add a social premium to the value of a barrel of oil. — Impose a tariff on imported oil. Phase tariff in gradually over two or three years.

SUMMARY AND CONCLUSIONS

We have shown in this chapter that energy price shocks can affect macroeconomic variables through a variety of different channels. Some of these effects are the result of rigidities in wages or prices, and in some cases the direction of the effect is ambiguous. We have also reviewed the implications that these effects have for economic and energy policy.

This chapter has covered a wide territory, and a brief summary of the main points and policy conclusions would be helpful. Such a summary is provided in Table 5-2.

Our understanding of the macroeconomic effects of an energy price shock is still incomplete. For example, the general equilibrium effects of an energy price increase on the real wage and on nonenergy prices is ambiguous and requires further empirical study. Also, this paper has ignored exchange rate effects, which preliminary studies have shown could be important.[11] Nonetheless, our current understanding of these macroeconomic effects has strong and robust implications for economic and energy policy. We feel that the policies outlined in the preceding section would help limit the impact of any future shocks. That impact will still be significant, but it should be manageable. The United States and other energy-dependent countries must avoid exacerbating the effects of energy shocks with misguided policies.

NOTES TO CHAPTER 5

1. For a discussion of problems associated with predicting oil prices, see Pindyck (1982).
2. Likewise, it is impossible for the Arab members of OPEC to create an energy shortage by imposing an oil embargo. The oil-producing countries can determine the quantity of oil they produce, but they cannot determine where that oil will ultimately be shipped, and this makes it impossible for an embargo to be effective against any single country or group of countries. An embargo can result only in production cutbacks, thereby reducing the oil available to all importing countries and driving up the price of oil until demand falls. The problem arises from the sharp increase in prices, not from the possibility that oil might be unavailable at any price.

3. This is discussed in Hall and Pindyck (1981a). For estimates of residential, industrial, and transportation energy demand elasticities in the United States and elsewhere, see Pindyck (1979).

4. Actually, GNP is a measure of domestic value added, that is, the net output of the domestic economy. Therefore, if energy is imported, total gross output, which includes the value of imported energy, may fall even if GNP (which measures the productivity of domestic factors) is unchanged.

5. This example is due to Nordhaus (1980). For general discussions of the relationship of energy to output and productivity, see Berndt (1980) and Nordhaus (1980).

6. This is the case in the model of Blinder (1981).

7. Hall and Pindyck (1981a, 1981b) assume this latter effect occurs. The ambiguity can be explained as follows: The change in nonenergy prices depends on the size of the fall in real income, on the sensitivity of the demand for real money balances to changes in real income, and on the weight of the price of energy in the price level. These factors are partially independent of those that determine the fall in real wages required by the direct effects discussed in the text. Therefore, it is possible that the fall in real wages that stems from the rigidity of nominal wages accompanied by the rise in nonenergy prices is larger than the fall in real wages that would prevail in the absence of adjustment problems.

8. Rotemberg (1983) shows conditions under which each of these cases will arise.

9. For further discussions of the macroeconomic effects of rising energy prices and the implications for macroeconomic policy, see Pindyck (1980), Hall and Pindyck (1981a, 1981b), Helliwell (1980), Rotemberg (1983), and Solow (1980). For a discussion of international differences in these effects, see Hall and Pindyck (1981a) and Fieleke (1981).

10. See Joskow and Pindyck (1979) and Pindyck (1981).

11. For a discussion of exchange rate effects, see Krugman (1983).

REFERENCES

Berndt, Ernst R. 1980. "Energy Price Increases and the Productivity Slowdown in U.S. Manufacturing." In *The Decline in Productivity Growth*, pp. 60–89. Boston: Federal Reserve Bank of Boston, Conference Series No. 22.

Blinder, Alan. 1981. "Monetary Accommodation of Supply Shocks Under Rational Expectations." *Journal of Money, Credit and Banking* 13, no. 4 (November): 425–38.

Branson, William H., and Julio J. Rotemberg. 1980. "International Adjustment with Wage Rigidity." *European Economic Review* 13, no. 3 (May): 309–32.

Economic Report of the President. 1982. Washington, D.C.: U.S. Government Printing Office.

Fieleke, Norman S. 1981. "Rising Oil Prices and the Industrial Countries." *New England Economic Review* (January/February): 17-28.

Hall, Robert E., and Robert S. Pindyck. 1981a. "Oil Shocks and Western Equilibrium." *Technology Review* 83, no. 6 (May/June): 32-40.

_____. 1981b. "What To Do When Energy Prices Rise Again." *The Public Interest*, no. 65 (Fall): 59-70.

Helliwell, John F. 1980. "The Stagflationary Effects of Higher Energy Prices in an Open Economy." Resources Paper No. 57. Department of Economics, University of British Columbia.

Joskow, Paul L., and Robert S. Pindyck. 1979. "Should the Government Subsidize Nonconventional Energy Supplies?" *Regulation* 3, no. 5 (September/October): 18-24.

Krugman, Paul. 1983. "Oil and the Dollar." In *Economic Interdependence and Flexible Exchange Rates*, edited by J. Bhandari and B. Putnam, pp. 179-190. Cambridge, Mass.: MIT Press.

Nordhaus, William D. 1980. "Policy Responses to the Productivity Slowdown." In *The Decline of Productivity Growth*, pp. 147-172. Boston: Federal Reserve Bank of Boston, Conference Series No. 22.

Pindyck, Robert S. 1979. *The Structure of World Energy Demand.* Cambridge, Mass.: MIT Press.

_____. 1980. "Energy Price Increases and Macroeconomic Policy." *Energy Journal* 1, no. 4 (October): 1-20.

_____. 1981. "Energy, Productivity, and the New U.S. Industrial Policy." In *Toward a New U.S. Industrial Policy*, edited by M.L. Wachter and S.M. Wachter, pp. 176-201. Philadelphia: University of Pennsylvania Press.

Pindyck, Robert S., and Julio J. Rotemberg. In Press 1983. "Dynamic Factor Demands and the Effects of Energy Price Shocks." *American Economic Review.*

Rotemberg, Julio J. In Press 1983. "Supply Shocks, Sticky Prices and Monetary Policy." *Journal of Money, Credit and Banking.*

Sachs, Jeffrey D. 1979. "Wages, Profits and Macroeconomic Adjustment in the 1970s: A Comparative Study." *Brookings Papers on Economic Activity*, no. 2: 269-319.

Solow, Robert M. 1980. "What To Do (Macroeconomically) When OPEC Comes." In *Rational Expectations and Economic Policy*, edited by S. Fischer, pp. 249-264. Chicago: University of Chicago Press.

6 TEMPORARY TAX REDUCTIONS AS RESPONSES TO OIL SHOCKS

R. Glenn Hubbard

The economic damage caused by the oil supply interruptions of the past decade has received considerable attention from both economists and policymakers. Arguments for government policy intervention center on the divergence between the private and social costs of imported oil and on the fluctuations in real national income and output that have accompanied the shocks. How best to mitigate the economic costs of a disruption remains an important, as yet unresolved, problem for policymakers.

This chapter focuses on the potential for temporary reductions in personal income taxes, or for the use of rebates, in mitigating short-run declines in aggregate demand associated with the oil shocks. Unfortunately, many analyses have referred to such a policy initiative under the name of "revenue recycling." Despite the fears of "fiscal drag" during an oil supply disruption—because the federal tax system is not neutral with respect to inflation and because of increased revenue from oil excise and profits taxes—sudden oil price increases are far more likely to raise the federal deficit, because of the reduction in tax collections occasioned by the decline in national income and the "automatic stabilizers" in government spending. Tax reductions for the purpose of reversing some upward impetus to the budget surplus are not warranted.

Below, a brief review of the issues surrounding the effectiveness of income tax reduction as a policy response is presented. Results of studies using econometric models are presented to quantify the benefits of temporary and long-run tax reduction proposals.

EFFECTIVENESS OF TEMPORARY TAX CUTS

One must be precise in qualifying and quantifying the potential benefits and tradeoffs implicit in "accommodating" temporary policy changes. Two principal questions underlie this discussion. First, to what extent can temporary fiscal policy changes bring about the desired effects on aggregate demand? Second, how can a fiscal stimulus be structured to avoid "overheating" the economy? Any justification of a plan to reduce personal taxes during a shock (or use rebates) must therefore rely not on the need to "recycle" increased federal revenue, but rather on the desirability of bolstering consumer spending and thereby recovering some of the loss in aggregate demand. Given this analytical approach to the problem, the effectiveness of a tax cut or rebate depends on the extent to and speed at which it is spent. Two issues immediately surface: (1) whether the rebate is temporary or permanent and (2) whether households internalize the government's intertemporal budget constraint.

The first dimension evolves from the considerations of the "life cycle hypothesis" of saving and consumption, according to which households have expectations of their permanent (wage and non-wage) income and smooth their consumption path over time. Temporary fluctuations in income are much more likely to lead to fluctuations in saving, as the consumption path has already been "set" as the solution to the household's lifetime optimization problem. At the very least, temporary changes in disposable income should have a smaller effect on consumption than permanent changes. In the context of a tax rebate program, rebates that are explicitly temporary are likely to be saved.

On another level, the effectiveness of a rebate program in stimulating consumer spending depends on the extent to which households view the government's budget process as part of their own. In the most extreme version of this view, tax rebates—increases in the government budget deficit—will have no impact on consumption, as households will perceive the expected *future* tax liability and increase their saving to offset it.

One factor that is likely to work against these two potential criticisms of a temporary rebate program to stimulate consumer spending is the existence of "liquidity constraints." Intertemporal optimizing models of the consumption-saving decision rely on perfect capital markets and the ability of individuals to borrow funds in an elastic supply at the "market" interest rate. In reality, there are institutional restrictions on borrowing (collateral requirements, required debt-service ratios, etc.), so that at least part of the population consumes all of its disposable income and would consume still more on the margin, given an increase in disposable income. Thus, a rebate program will be more effective in bolstering aggregate demand if it is targeted at liquidity-constrained consumers. This provides a macroeconomic rationale for rebate programs aimed at assisting the poor, who are most likely to be liquidity constrained, in meeting higher energy costs.

Most empirical work focuses on consumer spending to analyze the effectiveness of a temporary rebate program in stimulating aggregate demand. Consider the following simple consumption function:

$$C_t = a + b Y_t^* + d W_{t-1} \ , \tag{6-1}$$

where C, Y^*, and W are real (per capita) consumption, permanent disposable income, and nonhuman wealth, respectively. Conceptually, it is easy to think of a distributed lag on disposable income (Y) as being a proxy for permanent income, so that:

$$C_t = a + \sum_{i=1}^{n} b_i Y_{t-i} + d W_{t-1} \ . \tag{6-2}$$

If all of a temporary tax rebate were considered by individuals to be an addition to permanent disposable income, than a one-quarter rebate of amount R would raise consumer spending by $b_o R$, since $\partial C_t / \partial Y_t = b_o$. Suppose, however, that only part of the rebate is considered an addition to permanent income. Then the impact of a temporary rebate on consumption is:

$$\frac{\partial C_t}{\partial R_{t-i}} = \lambda b_i + (1-\lambda)\gamma_i, \quad i = 1, \ldots, n \ , \tag{6-3}$$

where λ is the fraction of the rebate counted as an addition to permanent income, b_i is the structural coefficient on disposable income

(i periods ago) and γ_i is the marginal propensity to consume out of transitory income (received i periods ago).[1] The consumption function (6-2) may be rewritten as:

$$C_t = a + \sum_{i=1}^{n} b_i (Y_{t-i} + \lambda R_{t-i}) + (1-\lambda) \sum_{i=1}^{n} \gamma_i R_{t-i} + d W_{t-1} . \quad (6-4)$$

For simulation purposes, then, the next step is to put forth reasonable values of central parameters λ and the γ_i's. The empirical evidence in Blinder (1981) on the effectiveness of the 1975 rebate program found that temporary taxes still on the books are treated roughly as fifty-fifty combinations of transitory and permanent changes.[2] Within this framework, I selected two cases: (1) a rebate that is fully perceived as an addition to permanent income, and (2) a rebate of which 50 percent is considered an addition to permanent income.[3]

MODELING THE EFFECTIVENESS OF TEMPORARY TAX CUTS

To quantify the potential benefits of income tax reductions in offsetting some of the economic costs of oil supply shocks, I will refer to some results of econometric modeling efforts. A small econometric model of the U.S. economy was used to simulate the effectiveness of an explicitly temporary personal income tax reduction. Evidence on the benefits of longer term reductions comes from the preliminary results of the Stanford Energy Modeling Forum model comparison project, "Energy Price Shocks, Inflation, and Economic Activity" (Hickman and Huntington 1982).

The model is designed to quantify the short-term economic costs of oil supply disruptions and to pinpoint the general equilibrium impacts of policy responses. A core macroeconomic model with real and financial sectors is linked to a model of the world oil market. Solution of the models is fully simultaneous and is accomplished through iteration. The government has at its disposal a set of fiscal and monetary policy instruments, with which it can influence aggregate demand and supply. The basic output of the model consists of a set of relevant oil prices accompanied by endogenous OPEC output projections and a set of macroeconomic variables dealing primarily with inflation, unemployment, financial variables, and income.[4]

The econometric model was used to simulate the costs of an oil supply disruption and the extent to which temporary tax reductions are an effective means of increasing demand. The simulations are done on a quarterly basis from 1982 through 1986 using information available through the end of 1981. To provide a basis for measuring the effects of an oil shock and to gauge the effectiveness of policy responses, I first constructed a control scenario: a state of the world without further oil supply disruptions or significant changes in policy.

In the control scenario, real GNP grows slowly (2.6%) in 1983, rather rapidly in 1984 and 1985 (4.9% and 4.5%, respectively); and moderately (3.6%) in 1986. The unemployment rate peaks in 1983 and falls gradually to 8 percent by 1986. Inflation remains between 6.3 percent and 7.6 percent. (This scenario is a basis for comparison and not a projection.) Oil prices continue to fall slightly until 1985, when they begin to rise gradually.

The disruption scenario represents a reduction in OPEC capacity of 7 million barrels per day during 1983. A disruption of this magnitude raises oil spot prices by as much as $35 per barrel and refiners' acquisition costs by as much as $24 per barrel. As expected, the disruption reduces economic growth and increases inflation and unemployment. The rate of GNP growth is reduced by about six-tenths of a percentage point in 1983 and 1984, and in 1985, GNP is reduced by about $20 billion by the disruption.

The next two scenarios represent the use of temporary personal tax cuts to reduce the costs of the disruption. I simulated the effects of a $30 billion personal income tax cut in 1983 under two different assumptions.[5] In the first case, "fully perceived," the temporary tax cut is assumed to have the same impact as a permanent tax cut of equal magnitude. As discussed earlier, agents in the economy do not behave according to the strict life cycle hypothesis and therefore do not simply save most of the tax cut they receive. In the second case, "50 percent perceived," the tax cut is perceived as being different from a permanent tax cut of the same magnitude. More specifically, the temporary tax cut is assumed to have the same effects as a permanent tax cut half its size, since many agents behave according to the life cycle hypothesis and save much of their tax cuts.

The fully perceived and 50 percent perceived tax cuts both diminish the costs of the disruption, particularly during the year of the disruption (and tax cut) without significantly affecting the price level. The tax cuts work primarily through their stimulus to consumption.

In fact, investment is ultimately reduced because of the tax cut. By the end of the simulation interval, the stimulative effects of the tax cut essentially fade away, and real GNP is only slightly higher than in the disruption scenario with no policy response.

As one would expect, the fully perceived tax cut is more effective in reducing the costs of the disruption than is the 50 percent perceived tax cut. In the final quarter of the disruption, the fully perceived tax cut reduces the GNP loss from $9.8 billion to $4.6 billion, while the 50 percent perceived tax cut only succeeds in reducing the loss to $7.7 billion. The stimulative effects of the tax cut are diminished by the upward pressure it exerts on interest rates through the budget deficit. In the final quarter of the tax cut, the long-term interest rate is 0.5 percent higher because of the tax cut. This accounts for the slight reduction in investment.

The ongoing study by the Stanford Energy Modeling Forum of the macroeconomic impacts of energy price shocks has examined the effects of a reduction in personal income taxes on income, unemployment, and inflation. Specifically, as a policy response to their "shock case" of a permanent 50 percent increase in the real price of oil, the participating modelers analyzed a permanent reduction in income tax rates of 10 percent (designed to provide a stimulus of approximately $30 billion).[6]

Results of the various models were mixed. Almost all models showed gains in the growth rate of real income and short-run reduction in the unemployment rate. In the long run, the tax reduction left the economic growth rate unchanged, but the composition of private spending shifted toward more consumption and less business fixed investment—principally because of the increase in interest rates associated with the widening public sector borrowing requirement. Because the model simulations of the impact of an oil shock in the absence of policy showed that the effects of an oil price increase— even a permanent oil price increase—on the growth rate of real GNP are likely to be transitory, temporary policy changes would be the more logical choice, provided that they can, in fact, achieve the desired effect on spending.

CONCLUSION

The short-run reductions in aggregate demand that have accompanied the sharp increase in the price of imported oil have led many economists and policymakers to advocate tax reductions as a counter-cyclical measure. Tax rebate or reduction schemes can certainly be constructed to increase the potential for consumer spending and to address certain equity goals. The effectiveness of temporary tax cuts in stimulating demand depends on the way in which agents determine their consumption decisions and on the extent to which they can attain their desired level of spending—as opposed to being rationed by liquidity constraints. The empirical results presented here indicate that those differences in behavior can produce very different results for the effectiveness of the policy. Under plausible parameter values, though, substantial benefits can be obtained from using temporary tax cuts at the onset of an oil shock. In reality, policy changes do not occur in isolation, and the ultimate test of the effectiveness of the temporary tax cut proposals depends on the total stance of fiscal and monetary policy.[7]

NOTES TO CHAPTER 6

1. Note that this approach is also consistent with a model of life cycle consumption behavior in the presence of liquidity constraints. In such a world, an explicitly temporary rebate would be saved, but the fraction of the population that is liquidity constrained would consume all, or at least a large part, of the increase in disposable income.

2. The empirical work in this area has by no means produced a clear consensus on the effectiveness of temporary tax changes in influencing consumer spending. For other studies of the problem, see Okun (1970), Hall and Mishkin (1981), and Hayashi (1981).

3. In terms of (6-4), assume that the propensity to consume out of transitory income is zero. Then, in case (1), $\lambda = 1$, and in case (2), $\lambda = 0$.

4. See Appendix A for more detail.

5. A program of this size can probably address the equity concerns highlighted by the oil shock. The program was fashioned as follows:

    First, the *Statistics of Income* of the Internal Revenue Service were used to determine the number of returns filed and taxes paid by tax bracket in 1978. Adjusting for inflation yielded a set of income brackets and corresponding average taxes in 1982 dollars.

    The rebate structure is based loosely on the 1975 rebate of a portion of 1974 income taxes—a rebate system that helped to end the 1974–75 recession. Everyone who filed a tax return in the year before the disruption begins would receive a rebate equal to 10 percent of the income tax paid in that year, unless taxes paid were more than $2,000 (in which case they would receive $200) or more than $5,000 (in which case they would receive a maximum payment of $500). Heads of poor households may "enroll" in the rebate system simply by filing a tax return, even if they owed no taxes. That would allow our simple tax rebate system to substitute for complicated changes in Aid to Families with Dependent Children and other welfare programs in reducing the costs to the poor of oil shocks.

The rationale for allowing rebates as high as $500—as opposed to restricting them to $200 lump sum rebates—is a desire to alleviate the effects of the oil shock on the automobile and housing industries. Reasonably high rebates to middle-income households may help to prevent a large drop in the demand for consumer durables.

6. For a description of the scenarios and the models involved, see Hickman and Huntington (1982).

7. For example, consider the combination of a temporary reduction in personal income tax rates with a temporary increase in the rate of growth of the money supply. The temporary tax cut is likely to provide a short-term stimulus to aggregate demand but is likely to have little long-term benefit. On the other hand, the benefits of monetary accomodation operate with relatively long lags, mitigating long-run reductions in capital spending.

REFERENCES

Blinder, Alan S. 1981. "Temporary Income Taxes and Consumer Spending." *Journal of Political Economy* 89, no. 1 (February): 26–53.

Hall, Robert E., and Frederic Mishkin. 1980. "The Sensitivity of Consumption to Transitory Income: Estimates from Panel Data on Households." National Bureau of Economic Research, Working Paper No. 505.

Hayashi, Fumio. 1982. "The Effects of Liquidity Constraints on Consumption: A Cross-Sectional Analysis." National Bureau of Economic Research, Working Paper No. 882.

Hickman, Bert, and Hillard G. Huntington. 1982. EMF 7 Study Design. Stanford University. Mimeo.

Hubbard, R. Glenn, and Robert C. Fry, Jr. 1982. "The Macroeconomic Impacts of Oil Supply Disruptions." Energy and Environmental Policy Discussion Paper Series E–81–07, Kennedy School of Government, Harvard University.

Okun, Arthur M. 1971. "The Personal Tax Surcharge and Consumer Demand, 1968–1970." *Brookings Papers on Economic Activity*, no. 1: 167–211.

7 ECONOMIC RESPONSE
Administrative Options and Analytical Framework

Michael C. Barth and Edwin Berk

The context of our discussion is a policy of allowing market forces to operate unhindered during a disruption in the supply of imported oil. However efficient the market may be in allocating oil supplies, unwanted consequences are likely to follow. Rapidly rising energy prices will cause a decrease in output and employment, individuals will experience personal hardships, and state and local governments may find their budgets under strain. Without abandoning a policy of market response to a disruption, the federal government may take direct action to respond to these adverse effects by injecting revenues into the economy.

This chapter explores one option for administering such an economic response and proposes a framework for the analysis of alternatives. The option explored combines a temporary reduction in the federal individual income tax and relatively unrestricted block grants to states.

Economic response measures are frequently referred to as "revenue recycling." Although we use the term "recycling" interchangeably with "economic response" in this paper, it is something of a misnomer. The source of any government revenues injected into the economy and, consequently, whether these revenues are being re-

The views expressed in this chapter are solely those of its authors and do not necessarily represent those of ICF Incorporated or any of its clients.

cycled may be unclear. This point requires some explanation before proceeding.

The overall economic effect of a sharp energy price rise may be a decline in the aggregate demand for goods and services. Consumers will devote a greater percentage of their incomes to energy goods, reducing their consumption of nonenergy goods and services. At the same time, the economy will be subjected to what is, in effect, a net increase in taxation. As oil prices climb, a greater fraction of the domestic spending flow will be siphoned off to foreign producers, much as if these producers had levied an excise tax on the United States. The profits earned by domestic oil producers will also climb, expanding the flow of tax dollars into the federal treasury through the corporate income and Windfall Profits taxes. Assuming that government spending does not automatically expand, the tax increase from both foreign and domestic sources would restrain the overall demand for goods and services. The result for the economy would be depressed levels of output and employment. The result for the federal government may be either a net increase in revenues or a net decrease, depending on whether the increase in income and excise taxes from the petroleum industry exceeds the reduction in taxes from industries hurt by the disruption.

From the standpoint of consumers, higher energy prices will mean personal hardships, especially for those in the lower income brackets, who spend a greater percentage of income on energy. Energy consumption or the consumption of nonenergy goods and services, or both, will have to be reduced. Moreover, basic services provided by state and local governments may be curtailed as these governments cope with the strain of higher energy prices on their own budgets.

The federal government may mitigate the possible contractionary macroeconomic effects caused by a disruption and alleviate the personal hardships by injecting revenues into the economy. If the federal government's net tax revenues increase, the increment may be returned to the economy, that is, recycled. If net tax revenues decrease, economic response measures will entail deficit financing. Thus, in the event of an oil supply disruption, the federal government can rely upon market forces to adjust prices and allocate supplies, while taking direct action through recycling to counter adverse effects on consumers.

Revenue recycling and economic response are topics of vigorous discussion, but the structure and administration of a viable program has been only superficially examined. The attention needed is long overdue. Hearings were held in December 1981 by a subcommittee of the Senate Finance Committee on legislation to provide standby revenue recycling authority for oil supply disruptions (U.S. Senate 1981). The legislation was not enacted, but there is little question that similar measures will be introduced in the future.

The present administration is firmly committed to the use of the market to regulate energy prices and supplies, having rejected the offer of discretionary authority to allocate supplies in an emergency. Yet, it is very likely that there will be public demand for the federal government to take some action in the event of a severe disruption. Economic response measures would enable the government to act directly in response to an energy emergency without interfering with the market's adjustment of prices and supplies. The idea of recycling federal revenues also appears in proposals to impose a tax on crude or refined petroleum and then rebate the revenues.

Our purpose in this chapter is to advance the analysis of feasible mechanisms for administering economic response measures. The analytical framework is presented in the following section. One option for an economic response program is then discussed in some detail, followed by a review of several alternatives.[1]

A FRAMEWORK FOR ANALYZING ECONOMIC RESPONSE MEASURES

The following analytical framework consists of seven factors that should be considered in evaluating proposed mechanisms for administering an economic response. These factors set forth a range of policy objectives that can be embodied in a response program. Some of these policy objectives may be incompatible with one another. For example, administrative simplicity may conflict with the objective of equity and broad coverage.

Underlying this analytical framework is the recognition that the appraisal of economic response measures rests upon normative policy considerations and that, in fact, the choice of an administrative mechanism represents a policy decision. Clearly, normative policy decisions must be made if empirical analyses or projections of impacts are to be useful. Thus, information about the percentage of the population covered by various mechanisms is useful in evaluating these mechanisms only within the context of a discussion of the objectives of an economic response program. The choice of an economic response mechanism that covers virtually the entire population, but at the expense of being administratively cumbersome, represents a clear decision about policy objectives. Our analytical framework, comprising seven factors, should contribute to discussions of normative policy considerations by providing a structure and encouraging explicit and serious review.

Speed of Implementation

The speed of implementation will depend not only on the complexity of the economic response mechanism but also upon the amount of planning, the resources available to administer the program, the degree to which administration is either centralized or dispersed among a number of organizations, and other such factors. It is considered desirable for an economic response program to return federal tax dollars to the economy as rapidly as possible in order to mitigate quickly the adverse effects of a disruption.

Preimplementation Requirements

Preimplementation involves the steps necessary to have an economic response program ready to operate when a disruption occurs. Some administrative mechanisms permit thorough preimplementation; others are difficult to prepare for in advance of implementation. The resource requirements can vary from large to minimal. Preimplementation may require the involvement of many agencies at several levels of government, or it can be conducted entirely by a small staff in one office.

Ease of Dismantling

An economic response program is temporary, to be instituted in narrowly defined circumstances for a specific purpose and then dismantled. Thus, it is important to consider the ease—the speed, the cost, the complications—with which a mechanism can be dismantled.

Systems Requirements

It is important that administrative mechanisms—whatever their requirements for data processing and financial management systems—can build upon systems already in place, avoiding the cost and complexity of designing and running an entirely new system. For this reason, special consideration must be given to the ability to use relevant existing data systems, such as the Social Security Administration's (SSA) Master Beneficiary Record and Earnings Reference File. Some alternatives can be based almost entirely upon an existing operating program, such as the federal individual income tax. An important systems-related consideration is whether several pre-existing systems must be merged. Merging disparate systems can be complicated.

Cost

Staff, facilities, and other administrative expenses are relevant factors in deciding among different mechanisms. The printing and mailing of checks, if a part of the economic response mechanism, may be a significant fraction of the total cost. If an existing administrative system is used, cost can be reckoned in terms of disruption to the program (if any) that normally uses the system. For example, if payments are made through the Social Security System, other aspects of SSA's operations may be slowed.

State, Local, and Private Sector Roles

Economic response measures can require minimal or significant involvement by state and local governments or the private sector. In some cases, this involvement is essential to the measure; in others, it is either an option or an unavoidable reaction, but not necessary to the federal government's administration of an economic response program. For example, recycling through reductions in federal income tax demands private sector involvement (employers have to adjust their payrolls); it may also require adjustments by states the income tax systems of which are tied to the federal system. The disadvantage of extensive state, local, or private involvement in an economic response measure is that substantial coordination would be required. On the other hand, decentralized administration may make it possible to target recycled funds more precisely to intended recipients.

Equity

Equity in an economic response program may be viewed in more than one way. It can be seen as ensuring that all citizens benefit from the program and few receive double benefits. Alternatively, it can be seen as the proportioning of benefits to the degree to which recipients are affected by an oil supply disruption. For example, a program may aim at concentrating benefits in regions the residents of which use oil heavily for home heating, or the program can aim at scaling benefits to individual household oil consumption.

Equity and broad coverage can be furthered by forming an economic response program from a combination of measures, each administered by different mechanisms. In so doing, it is important that the separate measures be compatible in timing. That is, the lead time for implementing the measures should be roughly the same and they should deliver funds concurrently. In addition, the measures should be selected to avoid overlaps and gaps in coverage. Different measures should not inadvertently provide benefits to the same recipients, nor should deserving persons be excluded from benefits.

Tradeoffs among these factors are inevitable. We describe two of the more important tradeoffs here; some others are suggested in the preceding descriptions.

Equity versus Administrative Cost and Complexity

Ordinarily, administrative cost and complexity increase if an economic response program is to achieve more complete coverage without overlaps or be able to target payments more accurately. A combination of separately administered mechanisms may be needed to expand coverage. Decentralized administration and a heavy reliance on staff-intensive support may be required to target recycled funds better. To obtain marginal improvements in either coverage or targeting, information requirements may grow dramatically.

Speed of Implementation versus Emphasis on Preimplementation

Extensive preparation can reduce the lead time necessary to implement an economic response mechanism. While rapid implementation is usually desirable, it may also be desirable to minimize preimplementation activities in order to limit up-front costs and maintain a low profile until an emergency occurs. As is usually the case, this tradeoff can be avoided by sacrifices in other areas. Rapid implementation can be achieved with limited preimplementation—as well as low administrative cost—by using a mechanism (such as the federal income tax system) that may have substantial gaps in coverage. In this regard, the concept of rough justice may be seen as a norm in evaluating recycling programs: Given the goals of speed and simplicity, is reasonable fairness and equity achieved? Even though added complexity or time could purchase greater equity, a roughly just program may be sufficient.

OPTIONS FOR ADMINISTERING AN ECONOMIC RESPONSE PROGRAM

This section outlines one option for an economic response program and then sketches several alternatives. The alternatives considered attempt to provide benefits to a progressively greater percentage of the population.

The option consists of a combination of two economic response measures: (1) a temporary reduction in liability for the federal individual income tax, implemented by means of an immediate reduction in withholding rates; and (2) block grants to states, with minimal restrictions on use. These two measures are complementary. A reduction in withholding for the federal income tax is a relatively simple and fast way to provide benefits to a large fraction of the population. It cannot, however, reach all of the population; in particular, some of those likely to be most disadvantaged by an oil supply disruption's effects do not participate in the federal income tax system. Block grants could be used by states to provide benefits to individuals left untouched by the tax reduction. Block grant funds could also be used to offset added costs to state and local governments resulting either from higher energy prices or from increased demands for basic and emergency services during a disruption. The federal funds available for an economic response program could be divided between these two component measures in any ratio desired, although one might expect a substantially larger percentage of funds to be directed to the tax reduction.

Administration of the tax reduction may be divided into three stages: preimplementation, implementation, and dismantling. Preimplementation involves the formation of an interagency group to formulate policy and procedures for the economic response program. Although it is unlikely that Congress would be willing to grant authority for a temporary tax reduction in advance of an imminent disruption, all necessary legislative materials could be drafted under the guidance of this interagency group.

Implementation of the tax reduction would require employers to reduce the percentage of their employees' salaries withheld for the federal income tax. Following declaration of an energy emergency, authorizing legislation would be introduced. On the basis of guidance developed by the interagency policy group, the Treasury Department—which would be responsible for the mechanics of implementation—would design new rate structures and print revised forms for employers and individuals. Employers would receive notification and new forms by mail and then make appropriate payroll adjustments. Self-employed persons would pay reduced taxes at quarterly filing times. If the disruption continued for an extended period, further changes could be made in tax liability and withholding rates, based upon the flow of federal tax revenues.

Dismantling procedures would be similar to those required to implement the tax reduction except that they would eliminate the reduction in tax liability. Data gathering and evaluation would continue throughout the program.

The block grants would provide federal funds for states to use at their discretion. States, however, would be expected to use their block grants to target cash benefits to individuals who do not participate in the withholding system by means of state-administered programs such as Low-Income Home Energy Assistance, Aid to Families with Dependent Children, or Food Stamps. The block grants could also be used to provide crisis assistance (e.g., shelter and emergency heating oil) to those who suffer extraordinary hardships because of an oil supply disruption. In addition, they could be used, for example, to provide funds to maintain mass transit services. More generally, block grant funds could offset higher energy costs for state and local governments.

Such block grants could most readily be implemented by using the standard operating procedures for the block grants established under the Omnibus Reconciliation Act of 1981. A form of revenue sharing could be employed as an alternative mechanism. Lead responsibility for the block grants, in the former case, would be assumed by the Department of Health and Human Services. The federal government would not manage the grants.

Preimplementation for block grants and the tax reduction would be conducted simultaneously. States would be notified after federal policy was formulated so that they could begin to develop necessary legislation and preimplementation programs to use the block grant funds.

Implementation of the block grants would begin with the enactment of authorizing legislation and the promulgation of regulations to clarify administrative procedures. The Office of Management and Budget would apportion funds and the Treasury Department would issue a warrant enabling the funds to be used. States would be required to submit applications, but the applications would be reviewed only for formal compliance with the authorizing statute. The size of each state's block grant would be determined by a statutory allocation formula. Awards would be made on a quarterly basis; states would use letters of credit to draw down awarded funds.

The administration of programs to use the block grant funds, including the handling of complaints about eligibility or the size of benefits, would be the full responsibility of the states. Because each state would design and administer its own program, the use of the grants as well as the timing and distribution of benefits might vary considerably from state to state.

Funding to states would be terminated when the decision was made to dismantle the economic response program. Some state programs funded by the block grants could be concluded quickly (e.g., crisis assistance); others (e.g., mass transit subsidies) would need to be phased out.

We estimate that preimplementation of the temporary tax reduction outlined above would take a minimum of 69 days; implementation a minimum of 37 to 47 days for essentially mechanical activities, to which several days must be added for policy decisionmaking;[2] and dismantling a minimum of 25 to 35 days. Preimplementation of the block grants would take a minimum of 134 days, implementation 12 days for largely mechanical activities and several days for policy decisionmaking, and dismantling 18 days.

While the combination of these two measures would broaden the coverage of an economic response program, furthering equity objectives, there would still be gaps in coverage, in part because federal control over the various states' use of grant money would be minimal.[3] To achieve more complete coverage, and in particular, to ensure that the aged and the poor are not neglected when benefits are distributed, the two economic response measures described above can be supplemented with a third measure. This third measure would be payments to individuals through the Social Security (RSDI) and Supplementary Security Income (SSI) programs. The SSA's existing (and efficient) systems would be used to send checks directly to recipients. Although these direct payments would enable an economic response program to reach those who do not participate in either the federal income tax system or in state-administered income transfer programs, overlaps in the coverage of the three measures could be substantial, as indicated in Table 7-1. A substantial amount of preimplementation would be required to reduce these overlaps. RSDI recipients would need to be eliminated from the SSI list and eligibility for energy payments established. At the same time, a special data file, modeled on the SSA's BENDEX (Beneficiary Data Exchange) and SDX (State Data Exchange) files, would need to be developed and provided to states so that RSDI and SSI beneficiaries could be weeded from the list of recipients of payments from state programs.[4] Dismantling, however, would be simple. An important mitigating consideration is that the SSA is currently operating at capacity with respect to both its systems and its personnel.

An economic response program can achieve even more complete coverage by placing greater emphasis upon direct payments. For example, a more comprehensive option would consist of three measures: direct payments to individuals through the RSDI and SSI programs, as under the previously discussed option; direct payments to individuals through the federal income tax system; and strictly defined block grants to states. Table 7-2 provides tentative data on the population that could be reached by the combination of these measures.

Table 7-1. Gap/Overlap Matrix Based on Program-Supplied Data (*percentage and number of row program participants who are also current participants of column programs*).

	WH	ES	RSDI	SSI	FS	AFDC	UI
				(millions)			
1. Income Tax Withholding (WH) Number of tax filing units (1975)	100% 71.8						
2. Estimated Tax System (ES) Number of tax filing units (1975)		100% 4.4[a]					
3. Social Security (RSDI) Number of individuals (1980)			100% 35.3	6.5% 2.3			
4. Supplemental Security Income (SSI) Number of individuals (1980)			51% 2.1	100% 4.15			
5. Food Stamps (FS) Number of households[b]	18.6% 1.5[c]		21.5% 1.7	21.5% 1.7	100% 7.9	42.6% 3.4	
6. Aid to Families with Dependent Children (AFDC) Number of families (1977)[d]	13% .458		3.6% .127	0% 0	73.5% 2.6	100% 3.5	.6%[e] .022
7. Unemployment Insurance (UI) Number of individuals (1980)[f]						.5% .022	100% 4.0

a. Lower bound.
b. Percentages are based on 1978 data. They have been applied to the current total participants for August 1980.
c. Upper bound.
d. The difference in years represented explains the difference between data entered for AFDC and for FS.
e. Data for 1979.
f. Approximately 5 percent of UI recipients are on some form of welfare.

Source: ICF Incorporated (1981a).

Table 7-2. Energy Payments: Coverage and Overlaps.

	(A) Total Number of Participants (millions)	(B) Percentage of U.S. Population ((A) ÷ 225 million)	(C) Number Reached by Higher Program (millions)	(D) Number Receiving Payment Through Program ((A) − (C)) (millions)	(E) Percentage of U.S. Population ((D) ÷ 225 million)
RSDI	35.3[a]	15.7	0	35.3	15.7
SSI	4.15[a]	1.8	2.3[a]	1.8	.8
Income tax withholding and estimated taxes	173.2	77.0[b]	11.6[c]	161.6	71.8
State-administered programs				26.3[d]	11.7
UI	4.1[e]	1.8	?		
FS	22.0[f]	9.8	?		
AFDC	10.7[g]	4.8	?		
"open window"	?				

Note: This table shows the total number of persons who are participants in each program and estimates of the number who are covered by programs higher in the hierarchy (listed in the left-hand column). The difference is an estimate of the number of persons who receive the energy payment from each program. All data are preliminary and tentative.

a. Social Security Administration statistics.
b. Estimated by Allen H. Lerman, Office of Tax Analysis, U.S. Department of the Treasury.
c. One-third of RSDI total participation. Estimated by Allen H. Lerman.
d. 225 million minus the sum of other (D) column entries.
e. Unemployment Insurance program statistics.
f. Food Stamps program statistics.
g. AFDC program statistics.
Source: ICF Incorporated (1981a).

The fullest development of the direct payments approach is presented in Chapter 8, in which Steven Kelman has analyzed and developed a program that would aim for complete population coverage by sending a simple lump sum payment to all nonminor individuals who satisfy statutory eligibility requirements.

Kelman estimates that it would be possible, by this means, to provide payments to almost 100 percent of the U.S. population over eighteen years of age. With respect to breadth of coverage, such a direct payment program would, if successful, achieve a high degree of equity. Equity in the context of economic response measures, however, may also be regarded as a matter of targeting benefits to those most directly affected by a disruption or those most in need of assistance, as noted above. Since almost all nonminor individuals are eligible for benefits, and the payment received by every eligible individual is identical, less is achieved in the way of this latter sense of equity. It is for this reason that 5 percent of the recycled funds would be reserved for block grants to states. Kelman notes that it would be possible under this particular scheme to target payments, say, by region or income, but targeting would delay the check issuance process and greatly increase the complexity of the program.

The likelihood of successfully establishing a program that could achieve 100 percent coverage and, thus, a high degree of equity in the former sense is itself far from clear. There is ample opportunity for political tampering in the implementation of this direct payments program. Eligibility must be determined by Congress; thereafter, the list of those eligible is compiled from the beneficiaries of a mix of federal and state-administered programs. To achieve complete coverage, this program would require perhaps an unrealistic degree of cooperation among the various federal and state bureaucracies with a role in assembling the list of eligibles. Those who think they bear the brunt of a disruption's effects are unlikely to rest content with a program that does not give special attention to their interests. The result may be the introduction of distortions into the careful balance on which the scheme depends to achieve universality of coverage.

Although Kelman estimates that implementation of this economic response measure could be completed within three months, this requires twelve to eighteen months of preimplementation. A combination of a temporary tax reduction and block grants, in contrast, might require no more than five months to prepare—probably even less if a crisis were on the horizon.

The option outlined in this chapter, tax reduction and block grants, falls far short of Kelman's direct payments program in potential breadth of coverage. While the emphasis on block grants permits greater targeting of benefits than with lump sum direct payments, this option generally settles for rough justice instead of full equity. But by so doing, the option remains simple and thereby lessens political vulnerability. Moreover, the administrative mechanisms used by these economic response measures are already in operation. Tax rates are altered quite frequently; the block grants now administered by the federal government are scarcely different in form from economic response block grants. Thus, there is good reason to think these economic response measures could be successfully implemented when needed.

CONCLUSION

The analytical framework presented in this chapter sets forth a number of factors that may be weighed in making decisions. While the framework may appear complex, it should simplify the evaluation of alternatives for economic response measures, providing a rational basis for choosing the alternative that is, if not perfect, then the least imperfect.

For example, the framework helps identify potentially conflicting policy objectives. One proposal for an economic response program might emphasize equity with respect to the percentage of the population reached or the degree to which benefits are related to hardships actually incurred. Another might emphasize the rapid injection of funds into the economy rather than mimicking an income transfer program; it would be aimed at offsetting income losses in general by restoring output and employment to the benefit of all citizens. The option offered in this chapter meets the second objective better than the first. While this option includes provision for reasonable coverage, it sacrifices precision and universality for the sake of feasibility and political robustness.

NOTES TO CHAPTER 7

1. Our analysis is based on work performed by ICF Incorporated under contract to the U.S. Department of Energy, issued in the form of two analytic reports and two standby operations manuals for economic response measures. See ICF Incorporated (1981a, 1981b, 1982a, 1982b).
2. "Policy decisionmaking" refers to all of the activities associated with deciding to start up the economic response, including executive branch analysis and consultation, congressional consultations, and the like. In our judgment, these activities can be compressed into a very short period of time in a crisis. It is, however, difficult to demonstrate this point. See the discussions in the fourth chapters of ICF Incorporated (1982a, 1982b).
3. The temporary tax reduction is subject to special administrative problems if initiated between September and December. It would be too late by these months for the Internal Revenue Service to modify forms and tables for the annual income tax return (used to reconcile withholding with actual liability). But if unmodified forms are used, a withholding rate reduction would be partly nullified when taxpayers filed their returns. This problem can be dealt with, although at the cost of some complexity and added administrative burden. See ICF Incorporated (1981a: 11-2, 1982a: 3-7, 3-8).
4. For a description of these systems and a discussion of systems problems, see ICF Incorporated (1981a).

REFERENCES

ICF Incorporated. 1981a. "Mechanisms for Recycling Federal Tax Revenues to Individuals and Households in the Event of a Sudden Increase in the Price of Oil." Prepared for the Office of Oil Supply Security, U.S. Department of Energy.

_____. 1981b. "Responding to Hardships from Oil Supply Disruptions: Analysis and Administrative Issues." Prepared for the Office of Energy Contingency Planning and the Office of Policy, Planning, and Analysis, U.S. Department of Energy.

_____. 1982a. "Operations Manual for Standby Temporary Tax Reduction During Energy Emergencies." Prepared for the Office of Energy Emergencies, U.S. Department of Energy.

_____. 1982b. "Operations Manual for Standby Block Grants to States During Energy Emergencies." Prepared for the Office of Energy Emergencies, U.S. Department of Energy.

U.S. Senate. 1981. "Standby Recycling Authority to Deal with Petroleum Supply Disruptions." Hearing Before the Subcommittee on Energy and Agricultural Taxation of the Committee on Finance, on S. 1354. 97th Congress, 1st session.

8 REVENUE RECYCLING USING CHECK WRITING

Steven Kelman

This chapter outlines an implementation plan for accomplishing revenue recycling. It is based on a report submitted to the U.S. Department of Energy (DOE) (Kelman 1982), although for reasons of space, a great many of the detailed implementation problems are not discussed here. The plan presented calls for the government, in the event of an emergency, to recycle funds by sending a check to all adult Americans. The implementation challenge is to show that this cannot only be done but done expeditiously.

The conclusion of the report to DOE was that getting checks to people was something that could be done. Using a modest number of already available lists, it would be possible to send checks to virtually every nonminor individual who is currently working, unemployed, retired, disabled, or receiving Aid to Families with Dependent Children (AFDC)—accounting for almost 100 percent of the U.S. population over eighteen. It is likely that the checks could all be mailed out to eligible individuals within three months or less of the decision to proceed with the program at the time of an emergency.

The program would be run by the Internal Revenue Service (IRS), with checks printed by the Bureau of Government Financial Operations of the Department of the Treasury, the normal check-produc-

Eugene Peters played an important role in the development of this chapter, and Seth Jaffe provided very helpful research assistance.

ing agency of the federal government. The involvement of other agencies would be limited to providing beneficiary computer tapes to the IRS. The IRS would create an "energy emergency list" of those eligible from its own Individual Master File (IMF) and from files provided by the Social Security Administration (SSA), the Office of Personnel Management (for civil service retirees), the Railroad Retirement Board (for recipients of railroad retirement pensions), and the fifty-one state programs (including the District of Columbia) administering AFDC. The energy emergency list would contain names, addresses, and social security numbers (SSN's) for all those eligible for payments. A "merge-and-match" computer program would rid the lists of duplication so that those eligible would not receive more than one payment. Computer tapes would be forwarded to the Bureau of Government Financial Operations for check production and mailing. Inquiries and complaints would be handled by IRS district offices. My recommendation was that one check be sent out to cover a one-year period, although it would probably be administratively feasible to send out a payment covering a six-month period and then repeat the payment six months later. Once the emergency was over and complaints dealing with nonreceipt of payments had been handled, provision should be made for destruction of the energy emergency list.

The program proposed appears feasible. The IRS has once before undertaken a similar program, when an income tax rebate was enacted during the Ford administration in 1975. Although the program proposed has never been tried in exactly this form, the plan builds on elements already in use and on a model that was used once before and worked. The program would be run by a single agency, reducing coordination problems dramatically. It would not require creation of any new agencies or institutions.

As much of the authorization for and shape of an energy emergency assistance program as possible should be dealt with by Congress and by the implementing agencies in advance of an emergency. I assume here that some sort of congressional authorization in advance of an emergency has allowed preparation of the contingency plans discussed below.

In the report to DOE, I recommended that the checks be a simple lump sum payment for all eligible individuals (double the lump sum for household units). I also recommended that a sum equal to 5 percent of the total funds to be recycled be made available to the states as untied block grants to deal with special hardships or problems as they see fit. It would be possible to target payments by region or by benefit receipt status, although targeting would delay somewhat the check issuance process and increase the number of complaints that must be handled after the checks had been sent out. Congress could also decide whether the payment was to be taxable. Making the payment taxable would increase its slant toward people with lower incomes.

The most potent objection to recycling has been that it simply could not be implemented successfully. As Chapter 1 made clear, in theory a recycling plan allows us to deal with the macroeconomic consequences of an energy emergency as well as responding to the equity issues raised by "letting the market work." However, recycling has heretofore not been given any real consideration as an option, for the simple reason that "it can't be done." It is easy to generalize about the benefits of recycling but daunting to think about what would actually have to be done in the event of an energy emergency.

There are quite a few obvious problems. Capacity would need to be quickly created to undertake a recycling program, which requires either establishing a new agency or agencies from scratch or dramatically expanding the capacity of some existing agency or agencies. How could one expect this to happen during the short period of time—a few months—available in order to get help to people expeditiously? A program that requires many people to apply for benefits would require significant personnel to process applications. A program paying benefits automatically must be able to handle complaints from people who claimed to be eligible but had not received payment—also a personnel-intensive task.

Withholding appears to offer a simple way out, but it entails major difficulties. Withholding misses huge numbers of Americans. Adding various agencies still leaves several significant problems. In particular, coordinating actions by the various agencies might prove to be an insuperable problem, both for ensuring that payments get made and for dealing with the inevitable complaints. In addition, such a recycling plan would still miss a large group: the self-employed.

On the other hand, opting for what would initially seem like the most logical way to recycle, namely to "send everybody a check," encounters the difficulty that a central registry with the names and addresses of the American people simply does not exist.

CRITERIA FOR EVALUATION

There are five basic criteria by which any recycling plan must be judged:

1. Equity: The point of the plan is to get money back to people and to eliminate legitimate cause for complaint about utilizing a market approach for dealing with the crisis. Thus, the first test that any plan must meet is that of coverage. Does the plan allow us to get money to everyone judged eligible? If not, who slips through the cracks and how can we deal with them? The decision to recycle money only to the poor, or to those with unusually high energy needs, or to some specific group (such as farmers), would obviate the need to worry about many of the issues that concern us here. Block grants to the states are the easiest course. Problems arise from the goal of reaching most, or all, adult Americans.

2. Flexibility: Because of the uncertainty inherent in a supply shock, the severity of the problem cannot be identified ahead of time. Hence, we cannot specify in advance how much money we will want to recycle. A program should be responsive to different funding levels. Also, any program should allow us to adjust payments for different groups (such as the poor or elderly) or areas of the country.

3. Speed and ease of implementation: Implementation problems can have very serious effects. Program delays will increase a supply disruption's economic costs as well as demands on the political system for additional action. Bureaucratic bungling—which becomes increasingly likely as program complexity increases—will not be treated mercifully by either the media or the public. This will be an important, visible program.

4. Costs: Any recycling program will have costs for development and implementation. There are also the larger costs (and benefits) to the economy of having the program in place and using it in the event of an emergency. Not all programs can promise the same economic benefits or costs, and an attempt must be made to reconcile the cost of implementation with the overall benefits to society.

5. Political acceptability: The program should be perceived as equitable and be able to meet the requirements of groups for whom the impact of an energy emergency are most severe. It should be able to deal with "ideological" concerns of some groups, such as issues of privacy and federalism.

DESCRIPTION OF THE RECYCLING PLAN

Implementing an energy emergency assistance program requires:

1. Establishing eligibility;
2. Establishing an energy emergency eligibility list;
3. Check production; and
4. Complaint handling.

Eligibility

Eligibility for energy emergency assistance payments would be established by Congress in authorizing legislation. The report to DOE recommended that Congress establish eligibility based on membership in at least one of the following categories:

- Individuals who have filed income tax returns for the year prior to an energy emergency;
- Nonminor individuals receiving a public retirement or disability pension;
- Individuals receiving AFDC; or
- Individuals holding a job for the first time during the year of an emergency.

Table 8-1 presents a list of sources of names for receipt of energy emergency assistance payments. This table includes estimates of the number of those appearing on each source list alone, as well as the number of individuals (eliminating overlaps) who would receive checks based on appearance on that list. Adding these lists together, it can be seen that approximately 101 million checks would be mailed, covering approximately 152 million individuals.

With the exception of the first-time job holders, names, addresses, and SSN's are available for all those in these eligibility categories, from a relatively small number of existing computer files, listed in Table 8-1. Almost all the names and addresses of those eligible will come from the federally available files of the IRS and the SSA. The only state-level files needed will be state AFDC files.

Table 8–1. Eligibility Categories for Energy Emergency Assistance.

Category	Source of List	Approximate Number of Participants for That Program Only (1980)	Approximate Number of Individuals Covered Based on Appearance on the List	Approximate Number of Checks to be Mailed Based on Appearance on the List in Question
Taxpayers	IRS Individual Master File	133.5 million	133.5 million	88.7 million
Social Security recipients (Title II RSDI)	SSA Master Beneficiary File	31 million[a]	15.5 million	9.7 million[b]
Supplementary Security Income recipients (Title XVI, SSI)	SSA Supplementary Security Record	3.9 million[a]	1.5 million	1.2 million[c]
Civil Service Pension recipients	OPM Annuity Role Master File	1.7 million[a]	245,000	245,000
Railroad Retirement Pension recipients	RRB Master Benefit File	1 million[a]	250,000	250,000
Welfare recipients (state AFDC programs)	State-level AFDC Files	2.2 million	1.2 million	1.2 million

a. Nonminor only.
b. Plus 4.8 million Electronic Funds Transfer payments.
c. Plus 130,000 Electronic Funds Transfer payments.

These eligibility criteria will cover the great majority of adult Americans. In particular, it should be noted that:

- They include those who are currently unemployed. Those currently receiving unemployment insurance will have worked during the previous year and thus filed tax returns during the previous year. They will therefore appear on IRS files.

- They include the working poor. The law requires that any wage earner have taxes withheld, even if at the end of the year that individual has earned an income so low that he or she will not be required to pay any taxes. Such individuals must file income tax returns in order to receive a refund for wages withheld.

- They include the self-employed. Since the self-employed file income tax returns at the end of the year, they will appear on IRS files from the previous year. (Withholding changes would miss the self-employed.) Nonworking spouses would also be included.

One group that would receive energy emergency assistance payments is high-school students who live at home but who have part-time jobs for which taxes are withheld. There is no way to separate them on IRS files from other people, since there is no information on a tax return about date of birth.

The following groups of nonminors would be excluded from a system that used the eligibility criteria noted above:

- College students who do not work at all;
- Nonminor dependents who are not receiving social security disability payments or Social Security Insurance (SSI);
- Those—mostly adult males—whose only source of income is county-level general assistance or food stamps;
- Divorced people whose only income comes from child support payments and who are receiving no social security or AFDC payments.

By far the largest group excluded would be nonworking college students, of whom there were approximately 3.9 million in 1980. The report to DOE recommended that these people should be ineligible for energy emergency assistance payments.

Those on county-level general assistance are generally people without children (and hence not eligible for AFDC) who do not work.

The number of such people is probably around 1 million, and they are extremely poor. The report to DOE recommended that a categorical grant, separate from any untied general block grant, equivalent to the number of general assistance recipients in each state times the size of the per capita energy emergency assistance payment, be given each state to be added on to general assistance payments.

Table 8–2 shows a calculation for how many people eighteen and older would be "missed" using these categories. According to the 1980 U.S. census, there were approximately 155 million Americans eighteen or older. Table 8–1 indicates (eliminating double counts) that checks would be sent to 152.2 million Americans. In calculating the percentage of the eighteen-and-older population missed, it is necessary to subtract the 3.9 million high-school students with part-time jobs who would receive checks but who are not counted in the eighteen-and-older category. It is also necessary to add three groups who have been accounted for but who do not appear on the lists: college students without any income (recommended to be ineligible), general assistance recipients, and first-time job holders (recommended to be eligible but receiving payments through grants to states and through special application during an emergency). These operations are performed in Table 8–3, which indicates that an estimated 155.4 million people have been accounted for. This figure is almost

Table 8–2. Who Would Be Missed?

Category	Number of Individuals	Proposed Procedures
College students who have not worked during the past year	3.9 million	Should be ineligible
General assistance/food stamps as only income	1 million	Per recipient categorical grants to states
First-time job holders who have never worked before	2.2 million	Should be eligible eligibility established at time of program
Nonminor dependents with no income or no income but food stamps, who do not qualify for SS Disability Payments	?, but small	

Table 8-3. Accounting for the Population Eighteen and Older.

Total estimated number of individuals to receive checks:	152.2 million
Subtract—number of high school students with part-time jobs:	3.9 million
	148.3 million
Add—college students who haven't worked over previous year:	3.9 million
General assistance recipients:	1.0 million
First-time job holders:	2.2 million
	155.4 million

embarrassingly close to the 155 million count in the census; I can only assure the reader that it emerged from several independent calculations not intended to lead to any predetermined result. Virtually 100 percent of Americans eighteen and older have been located. The program has essentially no gaps.

What about the size of the transfer? The report recommended that payment be a lump sum, determined at the beginning of an emergency by dividing the sum estimated to be available for recycling over a given time period by the estimated number of those eligible. Each eligible individual should receive a payment of that size. Household units that appear as such in the system, such as taxpayers filing jointly, would receive one check for double the lump sum.

Using information on the files that will make up the energy emergency list, it would be feasible, if Congress so decided, to target payments: (1) by region (state of residence); (2) by benefit receipt status (social security recipients, for example); and (3) by income (for taxpayers only). With the information on the files used to compile the energy emergency list, it would not be possible to target the payment by individuals' actual heating or driving behavior.

Administratively, targeting would slow down issuance of the checks and it would certainly complicate complaint handling dramatically, perhaps overwhelming the ability of the program to function effectively. In addition, the political conflicts engendered by any attempt at targeting might prevent Congress from taking the necessary planning steps before a crisis occurs.

An equal lump sum payment does not achieve complete equity. But a lump sum energy emergency assistance payment is far more equitable than providing no assistance at all in the event of an emergency, which is, unfortunately, a likely outcome given the political and organizational problems that a targeting attempt would engender.

Establishing an Energy Emergency Assistance List

A computer list needs to be generated with the names, addresses, and SSN's of those eligible for energy emergency assistance payments. Such a list can be used to issue checks to recipients. No central file of names and addresses exists in the United States, and therefore it will be necessary to create the list at the time of the emergency. Such a list can be assembled by the IRS National Computer Center, using existing files provided by the IRS, SSA, Office of Personnel Management, Railroad Retirement Board, and the fifty-one state programs that administer AFDC.

When an energy emergency is declared, the IRS would prepare a duplicate tape of names, addresses, and SSN's from its IMF. These tapes would be sent immediately to the Bureau of Government Financial Operations of the Department of the Treasury, which would begin sending out checks shortly thereafter. Also upon declaration of the emergency, the IRS would request that the other agencies send IRS tapes with the names, addresses and SSN's on their own files, in computer format according to IRS instructions. The National Computer Center would run a "merge-and-match" program for each incoming file against its IMF. Running such a program for a given incoming file will generate a tape containing names from that new file that had not appeared on the IMF. Such tapes would be sent to the Treasury for check writing as they become available.

The various tapes together will constitute the energy emergency assistance list. When completed, the list would be used to create a file, accessible on-line at IRS district offices, for complaint handling. When the emergency is over, the energy emergency assistance list would be destroyed.

IRS officials indicate that actual creation of an energy emergency assistance file, assuming that incoming tapes have the proper format, would not present any particular problems. The time, personnel, and computing resources required would not put great strains on their National Computer Center.

The potential source of difficulty involves format specifications for tapes that the other agencies would send the IRS. Incoming tapes cannot be run against the IMF unless record positions, blocking factors, and other computer format parameters follow IRS specifications exactly. The urgent nature of an energy emergency suggests that the IRS provide its specifications during the preimplementation stage and that the other agencies develop format software in advance, ready to be used in the event of an emergency. Some sort of annual updating would be helpful, and the expectation of having to spend time updating at the time of an emergency would be required. Realistically, it cannot be expected that the other agencies will have properly specified tapes "ready to go" as soon as an emergency is declared. There will be some delay between declaration of an emergency and the delivery of tapes from other agencies to the IRS.

As noted earlier, SSN's would be used as an identifier from the various data files. This means that there would be a problem any time an SSN is not a correct identifier of a person eligible for an energy emergency assistance payment. An emergency system cannot be expected to have lower levels of fraud than the subsystems (social security, etc.) on which it is based. But there are special problems created when files are matched.

Problems with incorrect and fraudulent numbers occur for several reasons. One source is simple transcription errors on a file. All the systems involved have strong incentives to catch transcription errors, given the importance of the SSN as an identifier, and they are continuously taking steps to do so. Problems arising from transcription are generally quickly resolved.

Fraudulent numbers may arise in two ways. First, an individual may fraudulently obtain an SSN from the SSA. There is no way that this could be detected. It obviously cannot be expected that an emergency program could improve on fraud detection in ongoing programs.

The second way is easier to detect. An individual, such as an illegal alien, may invent an SSN. If an individual uses the invented number when applying for benefits or filing a tax return, a computer check by the IRS or SSA will show a "no-match" for the four-letter name control associated with the SSN. This no-match will cause the entry to be placed on the invalid segment of IRS or SSA files. These people would not be eligible for energy emergency assistance payments, so that this type of fraud need not create a problem.

Finally, a legitimate SSN may be associated with two different names on different files. An unmarried woman may be named Mary Smith when she applies for an SSN at the beginning of her working career and have changed her name to Mary Jones on a joint tax return at the time of an energy emergency. In cases such as these, the merge-and-match will show two names associated with a given SSN. Such name inconsistencies frequently take some time to be detected and reconciled between the IRS and SSA.

The report to DOE recommended that where operation of the merge-and-match program indicates two different names for one SSN, payments be sent to both names, perhaps after a zip code check. It estimated that this would result in several hundred thousand duplicate checks being sent out. Sending duplicate checks is administratively easier than dealing with complaints from people who should have received checks but were excluded because of a matching error. In an emergency system such as this, with relatively modest, nonrecurring payments, it is better to err on the side of overinclusion than underinclusion.

The creation of an energy emergency assistance list would be the most complicated part of preimplementation because it involves many other agencies in addition to the IRS. At a minimum, the other agencies that would be providing tapes would need to know their role in an energy emergency assistance program, and one "pilot run" using the IMF and tapes submitted by the other agencies would need to be undertaken. The IRS should also annually update its tape format specifications for the other agencies.

During an emergency, the IMF should be sent over to the Treasury quickly, so that check production can begin. It is not unrealistic to expect that the list consisting of names and addresses from the SSA files not on the IMF can be sent to the Treasury within two months of the beginning of the program, which would mesh with the Treasury's check-printing schedule outlined below and be consistent with the goal of having all checks mailed out within three months. Some of the other lists, including some state AFDC lists, will probably come in rather quickly; others are likely to drag. The number of names involved here is not large, but the people involved are generally needy, and it is desirable to get them their payments as soon as possible.

Check Production

Checks must be printed and mailed to those eligible for an energy emergency assistance payment. The checks could be printed and mailed out of the eight regional disbursement centers of the Bureau of Government Financial Operations of the Department of the Treasury. This is the same bureau that normally prints federal government checks, including monthly social security and tax refund checks. Tapes would be delivered to the Treasury by the IRS. Using existing Treasury facilities, it would be possible to have all checks mailed out within three months or less of the decision to go ahead with an energy emergency assistance program. Checks would be mailed on an ongoing basis during those three months, so most recipients would have them much earlier. The first checks should go out within days after the program is triggered.

The IRS should send the relevant information from its IMF to the Treasury as soon as it becomes available, allowing the Treasury to begin check production and mailing days after the program is triggered. Tapes with additional names and addresses of eligible recipients can then be sent to the Treasury as they become available. A crucial point here is that the system can start up before all the tapes have come into the IRS. Otherwise, the whole system would stand hostage to laggard tapes, and it would be unlikely that checks could ever be issued quickly enough. Check writing can begin immediately if the IRS simply sends it own IMF off to the Treasury at the beginning of an emergency. Additional tapes would then be sent as they became available, but since most of the names for assistance payments are on the IMF anyway, the Treasury would have no "down time" waiting for the non-IRS tapes to arrive. To get checks to the elderly and to AFDC recipients out more quickly, the Treasury could be instructed to interrupt its check printing upon receipt of name and address tapes produced by the merge-and-match program with tapes from other agencies.

The processes of printing and mailing checks from computer tapes are identical to tasks already performed by the Bureau of Government Financial Operations, which mails out over 700 million federal government checks per year. This organization also has annual experience with peak loads, during tax season, when almost 70 million tax refund checks are mailed out over six months. In addition, the Treasury had a one-time experience in 1975 with sending out a large number of additional checks over a short period of time, with only a brief lead time. This occurred after passage of the Tax Reduction Act of 1975, which provided for tax rebates of up to $200 for every taxpayer. The 1975 tax rebate was an undertaking the total size of which was only slightly more modest, in terms of its demands on check-printing and mailing capabilities, than the proposed energy emergency assistance payment. The success in issuing tax rebates in 1975 is a reassuring indication that the Treasury's check-writing capability could handle the task of printing and mailing energy emergency assistance checks.

A total of 101 million checks would need to be printed. A proposed schedule for increasing check-printing capability during the period required to print these checks was developed in consultation with officials of the Bureau of Government Financial Operations. It involved (1) having existing personnel work overtime and (2) hiring temporary personnel to allow existing machines to be used more intensively. The schedule allows a (conservative) estimate that 100 million checks, over and above the normal work of the organization—which would continue during the same time—could be printed within three months.

Undeliverable checks, sent to people who have moved, will be a problem for the program and a major reason that people will complain about nonreceipt. At the beginning of an emergency, program staff at the IRS should contact the Postal Service to assure that the local post offices know that checks will be coming and that forwarding them when necessary is important.

Complaint Handling

The two largest categories of eligible nonrecipients would be (1) people who had moved since filing their most recent income tax return whose mail is undeliverable and (2) first-time job holders during the year of the emergency, who had not previously filed tax returns or received social welfare benefits. The main source of the first problem is that the IMF has correct addresses only for the most recent tax filing year. The further from tax filing season in a given twelve-month period an emergency occurs, the greater the undeliverable mail problem will be. In terms of first-time job holders, this problem will be greater the later in the calendar year an emergency occurs.

In addition to these predictable categories of nonrecipients, there will be people who deny, correctly or incorrectly, having received a payment even though one was indeed mailed to their current address. These claims may arise, for instance, because of stolen checks or cases of divorce where one spouse cashed a check due both jointly. They may involve fraudulent attempts to obtain duplicate payments. There will be claims from people to whom no check was mailed. Some of these people will be eligible and others ineligible.

Finally, as soon as the program is announced, and during the period that checks are being mailed, there will be inquiries from people with questions about eligibility or anxious that they have not received their check—especially when neighbors have. A complaint-handling function is needed to deal with such inquiries.

The Taxpayer Assistance Service (TAS) of the IRS can handle such complaints and inquiries by telephone and mail. Most of the work would be done through IRS district offices. The "complaint window" will not open until all checks have been sent out. At that time, individuals with inquiries or complaints would be directed to fill out special forms and mail them to the IRS or to call special telephone numbers that would connect them with the nearest IRS district office.

The TAS is accustomed to dealing with peak-load problems of the kind that an energy emergency assistance program would create. Peak loads occur during tax-filing season and, to a lesser extent, in the months thereafter. Each year, the TAS hires and trains temporary personnel to handle peak loads; it also purchases additional telephone circuits for incoming phone calls.

Personnel at the district offices have access to on-line data capability that allows them to access key information from the individuals's tax return. This information is retrieved from the IRS Integrated Data Retrieval System, which contains selected information from the IMF.

The TAS has standard operating procedures for dealing with complaints about refund checks. The largest source of problems comes from people who have moved and whose mail has not been forwarded. If callers have moved, and if their current address is different from the address on the "refund file," a new refund request form could be sent. A new check is issued, and a stop-payment order is placed on the old check. The complaint-handling process for those who deny having received an energy emergency assistance check would be virtually identical to processes already in use.

An important question is whether the IRS has the capacity to deal with the expected volume of complaints. The report estimated that, depending on the month the complaint window opens, there would

be between approximately 2.5 and 3.8 million complaints. The varia-
tion results from differential list staleness due to seasonality in address changes and the differential impact of first-time job holders,
given that school graduation in June produces a bulge in new job
seekers.

To see whether the IRS district offices could handle such a load,
it is necessary to compare this expected demand from energy emergency assistance complaints with the demand made on IRS district
office taxpayer assistance capability each year during peak-load tax
season. In particular, the district offices would need to expand their
telephone circuitry to handle the increased number of phone calls,
to enter information (name and SSN) from complainants, and to
access stored information regarding check receipt. This requires ter-
minals at the district offices and data communications capacity to
access the central computer.

The report to DOE examined in some detail the ability of the IRS
district offices to expand their capacity. The system appeared capable
of dealing with the predicted number of complaints, except possibly
for some problems with insufficient data terminal capacity. Various
contingency plans should be developed in case terminal capacity
proved to be a problem at some district offices or the number of
complaints turned out to be greater than estimated. These include
possible methods for channeling call-in times, encouraging mail
rather than telephone inquiries, and accessing data to respond to
inquiries in the evening only, rather than while the complainant is on
the telephone.

Several important steps need to be taken prior to an emergency to
prepare for establishment of a complaint-handling capability. These
include:

1. Statutory language authorizing override of civil service procedures for personnel to be hired to handle complaints;
2. Preparation of hiring estimates for temporary personnel and for
   emergency increases in telephone circuitry and terminal capacity;
3. Preparation of plans for handling complaints at the district offices, including plans for possible expansion of terminal capacity;
   use of off-hours processing of on-line requests, for reducing, if
   necessary, the number of complaints to be serviced over the tele-
   phones; and leveling out the flow of telephone calls;

4. Preparation of standard operating procedures for dealing with different types of complaints and of training manuals for newly hired personnel;

5. Determination of what will constitute proof of eligibility for people in categories who do not appear on the energy emergency list; and

6. Preparation of forms, with accompanying instructions, to be filled out by various categories of complainants.

Inevitably, at the time of an emergency, some people would immediately call IRS district offices with questions about eligibility, non-receipt of checks, and so forth, despite public announcements that checks would take three months to mail out. Capability is needed to deal with these immediate requests. Public information (press releases, media contact, etc.) should be tailored to reduce the number of immediate inquiries. Other agencies (DOE, SSA, the White House) and congressional staffs should be provided with basic information about the energy emergency assistance payments. During the crisis preparation period, additional telephone circuits should be ordered on a "rush" basis for district offices to accommodate the increasing volume of inquiries, with recorded intercept messages used until they arrive.

Attempts should be made to get as many of the first-time job holders as possible to fill out forms establishing their eligibility to ensure that they receive their checks.

The complaint window would not open until checks had been mailed out to everyone on the energy emergency assistance list. This means that there would probably be at least three months, which should be sufficient to organize the complaint-handling capability.

THE PROGRAM ON BALANCE: PROS AND CONS

The most important conclusion of this work is that recycling can work during an emergency—even a recycling program that sets the ambitious goal of helping all Americans. The keys to that ability include reliance on a single agency, the use of existing routines adapted to these new purposes, and the effort to keep the program simple by writing people a check. Both the IRS and the Bureau of

Government Financial Operations, which would print the checks, are accustomed to peak-load demands on their system, which makes it feasible to gear up during an emergency.

It is time to return to the criteria set forth earlier in this chapter. The merits and failings of this particular plan must be catalogued and then contrasted with the alternatives. Recall each of the five headings presented earlier:

1. Equity: If the decision is made to assist the American people en masse, check writing appears to be the best way to do it. Using computerized lists it is possible to identify and to distribute a check to almost every American. If, by contrast, the decision is made to target only a narrow portion of the population, the system outlined here is unnecessary.

2. Flexibility: The plan is flexible, functioning with a wide range of rebate sizes. But it creates the problem that differentiation beyond broad classifications is difficult. For example, it would be impossible to target some special classes of recipients (such as low-income consumers who heat with oil) for extra assistance using the lists available. The presence of addresses could enable some regional targeting of the program. Since the program is a type of grant, such a distinction may be politically feasible. On the other hand, it is unlikely that Congress would allow different withholding rates to exist in different regions. Also, the rebate could be made taxable, and therefore more money could be targeted to the less wealthy. Still, this lack of precision in the ability to target funds is a drawback.

It must be remembered that eligibility criteria beyond what is readily available on existing computer lists require an individualized application procedure, which is extremely personnel intensive and difficult to apply to large groups of applicants. Such a system also slows down the issuance of payments. A hybrid program, where an energy emergency assistance check, equal for all, is supplemented with a block grant to states larger than the 5 percent suggested, would allow states to develop larger targeted programs.

3. Speed and ease of implementation: The use of a single lead agency will reduce the problems of program coordination, and the presence of a responsive and flexible service bureau will mitigate frustrations with lost benefits or unforeseen problems. Unfortunately, even if extensive testing is carried out ahead of time, this is still a new program. However the tests turn out, there can be no guarantee that the actual crisis implementation will mirror them.

4. Costs: Resources were not available in developing this implementation plan to generate detailed cost estimates. The cost estimates totaled $55 million for program implementation. This figure is an underestimate in that it excludes certain IRS costs, costs to agencies other than the IRS and the Treasury, and periodic computer software updating. But this figure is not a dramatic underestimate. This cost appears to be modest for a program that might, if the price of oil doubled in an emergency and the entire incremental revenues from the Windfall Profits Tax were recycled, distribute $25 billion.

5. Political acceptability: The program has several characteristics that improve its political appeal. First, it includes virtually all adult Americans. Second, it would be highly visible in the event of a supply emergency. The appearance of a rebate check would help assure the public that the government was sincere in helping them cope with the shock of higher energy prices. Third, the phasing out of the program is automatic. The checks can be a one-shot effort. There will be no need for a "tax increase," as higher withholding rates are likely to be perceived, once the need for emergency action slackens. Lastly, the program relies largely on existing agency personnel for its implementation. It does not create a new bureaucracy.

The program has political liabilities as well. Notably, it is very centralized. However, the addition of block grant components can help broaden the program's political base. The development of a central list of Americans may conjure, among some, visions of *1984.* Because the information on the list is of the most rudimentary variety and would quickly be destroyed, such criticisms do not have much substance. Their appearance, however, is inevitable.

The biggest political problem of any revenue recycling plan is, of course, getting Congress to act in advance of an emergency. Without the pressure of deadlines the issue may become bogged down in debates on benefit targeting—particularly regional targeting. If Congress waits for an actual emergency to establish a program, the chances of getting assistance to people expeditiously are small, both because of the time delays in passing legislation and the inability to do the modest, but nonetheless necessary, preimplementation work this plan requires.

As was pointed out earlier, a significant argument against congressional action on preauthorization of revenue recycling has been that such a plan could not be implemented. It is our hope that by addressing implementation issues in advance, we can expand the menu of policy choices in the area of energy emergency planning that are available in the political system.

REFERENCE

Kelman, Steven. 1982. "Revenue Recycling: An Implementation Plan." Report to the U.S. Department of Energy, October 20, 1982. Also available as Harvard Energy Security Program Discussion Paper H-83-01.

STOCKPILE POLICY

9 INSTITUTIONAL ALTERNATIVES FOR FINANCING AND OPERATING THE STRATEGIC PETROLEUM RESERVE

James L. Plummer

SOME PROBLEMS FACING THE STRATEGIC PETROLEUM RESERVE

Financing

When the Strategic Petroleum Reserve (SPR) program was originally established in 1975, the goal was a 1 billion barrel reserve. Later, a plan was submitted to Congress for completion of a 750 million barrel reserve by the end of 1989, while retaining the 1 billion barrel level as a long-term goal. The fiscal year 1984 budget submitted to Congress in January 1983 implied attaining a level of only 600 million barrels by the end of 1989. This level would be below that considered desirable in many of the economic studies of optimal SPR size.[1] This further reduction would not be based on physical construction constraints, but rather on problems of budgetary stringency.

There is a history of financing problems for the SPR. The Carter administration took all money for SPR oil acquisition out of the budget during the 1980 election year and gave Saudi opposition to SPR filling as the excuse. Then, in 1981, Congress took most SPR

This chapter draws partially upon a report prepared for the Strategic Petroleum Reserve Office of the U.S. Department of Energy in April 1983. The author wishes to acknowledge the helpful comments by Howard Borgstrom and Jerry Temchin of the SPR Office.

funding out of the budget on the assumption that it could be financed via off-budget securities linked to the price of oil. When that idea turned out to be infeasible, the capital financing for the SPR was shifted to a special off-budget Treasury borrowing account, which remains the financing system today.

Even though the SPR enjoys considerable bipartisan support in the Congress, it often is treated as a marginal item in times of budgetary squeeze. Given this persistent problem, it is appropriate to consider alternative mechanisms of financing.

Drawdown

Section 161 (d) of the Energy Policy and Conservation Act (EPCA) of December 22, 1975 states that "no drawdown and distribution of the Reserve . . . may be made, unless the President has found that implementation . . . is required by a severe energy supply interruption or by obligation of the United States under the international energy program."

The history of the 1973–74 and 1979 interruptions, plus the experience with other federal commodity stockpiles, justifies skepticism about whether such a presidential declaration would be made early enough from the point of view of avoiding substantial losses from oil price increases. It is quite possible that, within the context of an international oil crisis, the timing of this decision could be viewed in terms of its impact on international politics rather than its impact on the U.S. economy. The president and his foreign policy advisers would be most concerned about not prematurely alarming our allies with a presidential declaration that a "severe energy supply interruption" existed. The decision about when to begin drawing down the SPR, or a limited part of the SPR, should not be made using national security or foreign policy criteria.[2] Yet, that is what is likely to happen if it is a highly visible presidential declaration made in the midst of an international crisis. An SPR drawdown decision made on economic criteria would precede by weeks or months one based on foreign policy criteria.

Method of Sale

There is a strong consensus among economists that the SPR should be drawn down in a way that is consistent with free market pricing.

The SPR drawdown should not become entangled in attempts to subsidize one group of refiners or another but should give them all the same access to SPR oil on the same market terms. The SPR drawdown plan announced in December 1982 is consistent with these sound economic principles and thus is strongly to be preferred over any system based on nonmarket allocations or controlled prices. However, many drawdown systems are consistent with free market principles, and some of them could be better than the spot market auction system in the Department of Energy (DOE) plan.[3] Some of these drawdown systems will be described below.

A REVIEW AND CRITIQUE OF PREVIOUS PROPOSALS

Bonds Linked to Oil Prices

In 1980 and 1981, the pressure to cut the federal budget created a legislative compulsion to find an off-budget financing instrument for the SPR. Hearings were held on many bills that would have created oil-price-linked securities. The support for this general idea crossed party lines. Among the sponsors of bills of this type were Senators McClure, Warner, and Domenici, and Representatives Dingell, Stockman, Jones, and Gramm. This chapter will not lay out the pros and cons of all the various bills, but it will outline the broad conclusions that emerged from the hearings.

First, it was concluded that being linked to the price of oil was not a sufficient attraction for long-term institutional investors.[4] If these securities were to be sold in large volumes they would have to bear interest, and the principal would have to be collateralized by either a federal guarantee or some other kind of impressive asset (like the oil in the SPR). Without attractive interest and collateral, the securities would trade in limited volume on commodity exchanges, where the time horizon is a year or less, rather than in bond markets—or, more precisely, convertible bond markets—where the time horizon is much longer. At the end of the hearings, it was clear that the oil price linkage feature was one kind of attraction that could be added to a bond of three- to ten-year term, but that feature would add to other interest and collateral characteristics rather than substituting for them.

Another broad conclusion that emerged from the hearings related to the so-called "automatic disruption redemption" feature. If a security had this feature, then an SPR drawdown and sale of a given dollar value of oil would trigger redemption of an equivalent dollar value of securities. This feature has macroeconomic advantages because it guarantees that an equal volume of money is injected into the economy via redemptions as is withdrawn from the economy in SPR sales. However, these advantages were thought to be outweighed by the disadvantage of giving a windfall to a "crisis constituency" of bondholders and enduring the political backlash from consumers of oil products. Besides, it is possible to have many of the same macroeconomic advantages with fixed-term securities and a policy that the money proceeds from SPR sales be invested in Treasury bonds until the time for SPR refill arrives.

A Quasi-Public SPR Corporation

Proposals have been made by Plummer (1982a, 1982b) and Alm and Krapels (1982) to create a quasi-public corporation that would include some of the *nonregulatory* features of the systems now in use by the Federal Republic of Germany, Switzerland, and the Netherlands. The main advantage of this kind of system is that it removes at least part of the financing requirements from both the private sector companies and the federal budget. The yearly debt service and operating expenses would be met by levying a "user fee" on crude input to refineries, plus imports of refined products.

There are many possible variations of debt that could be issued by such a public-private corporation. If its debt merely consisted totally of federally guaranteed bonds, then it would be only cosmetically different from the present system of using a special off-budget SPR borrowing account at the Treasury. However, it might issue part of its debt (say 50%) in the form of bonds that carried no federal guarantee but were collateralized by some of the oil in the SPR (or the proceeds of selling the oil in a disruption). These kinds of bonds would compete with corporate bonds rather than government bonds in capital markets and thus would not be just a "deficit by another name." An SPR corporation could also experiment with adding an oil-price-linkage feature to some of its debt instruments to see how much it increased their attractiveness. These financing alternatives will be considered below. The federal government would realize a sizable one-time reduction in the budget deficit.

These proposals have not generated much excitement. The general reaction seems to be that it might have been a superior financing and organizational scheme if it had been done in the mid-1970s but that the benefits of moving to such a system now may not be worth the organizational disruption to the current SPR program. This is a serious criticism, but three new factors have caused reexamination of this institutional alternative: (1) the new financing problems of the SPR, (2) the possibility that a quasi-public corporation might be combined with other mechanisms for improving SPR drawdown, and (3) greater receptivity to using the quasi-public SPR corporation for a small part of the SPR (not parts still under construction) rather than using it for all the SPR. These issues will be taken up later in this chapter.

Proposals of the Industrial Oil Consumer Group (IOCG)

Some very interesting proposals have been made by the IOCG.[5] The IOCG proposals address mainly the drawdown problems of the SPR.

1. In order to achieve prompt drawdown of the SPR as soon as an interruption occured, conditional sales would be held before an interruption.
2. Routinely, even when no disruption was in sight, sales of rights to purchase SPR oil during an emergency would be held.
3. Exercise of such rights would be conditional upon a presidential declaration of a disruption under EPCA.
4. Only a limited amount of oil (no more than 10% of the SPR) would be offered for sale.
5. All bidders at the auction would have executed agreements with DOE with definitive terms, so that the details of the transaction were agreed to and understood prior to the declaration of an emergency.
6. There would be a minimum bid price (e.g., the spot price plus 5 to 10%) so as to encourage private storage.
7. Once a presidential declaration was made and drawdown of the SPR began, a successful bidder would be able to exchange or transfer the oil or the rights to purchase the oil.

8. All bids would be given priority in terms of the price bid per barrel until the offered quantity of oil was all bid.

9. When drawdown is taking place, any oil not claimed by a successful bidder within a fixed period would be made available to the next highest bidder. Those not accepting oil would be subject to a penalty in the event of a default. The penalty would be set at a level that would reduce the prospects of speculative bids without raising it to a level that would force the companies to accept oil from the reserve if the price increases had moderated in the meantime. This partial penalty system places this contract somewhere in between a futures contract (which is strictly "take or pay") and an options contract (where the buyer can let his option expire with no further obligation).

10. If the interruption ended before the oil was delivered, the government could release buyers from their contracts if they wished to be released.

11. After the presidential declaration had been made and SPR drawdown begun, any new contracts for forward delivery would be made without penalty clauses. Thus, during a disruption, the contracts sold would be tantamount to options contracts.

The IOCG proposals were deliberately designed so that they could be implemented under existing laws. They are definitely an improvement over DOE's simple system of spot market auctions after a presidential declaration. The predisruption sale of conditional contracts for forward delivery of SPR oil allows the SPR to have a favorable influence on oil market prices long before a drop of oil ever flows out of the SPR.[6]

One potential problem with the IOCG proposals is that a bidder could bid any price for the oil so as to ensure access to the SPR. If international tensions built up considerably before a presidential declaration was made, then a few panicky bidders could bid high prices for a small amount of oil. When the results of the forward contract auction were announced, it is likely that the press would play up the highest prices bid, and this media attention could have an unfavorable "price leader" effect on the spot market.

The other problem with the IOCG proposal is the problem with the current law, as described above. Refiners would be bidding on these contracts on the expectation that the condition of the contract, a presidential declaration, would be made promptly as soon as spot prices jumped. Yet, for the reasons described earlier, such a presidential declaration would likely be deferred beyond that point because of foreign policy considerations. The refiners holding successful bids would become a political constituency pressuring for a quicker presidential declaration. Other things equal, it would be better to have such a political constituency than not to have it. However, it would be even better to have a system in which part of the SPR could be drawn down by subpresidential authority based on judgments of economic impact, rather than waiting for all of the SPR to be released when foreign policy events dictated. This alternative will be discussed below.

Predisruption Sale of SPR Options

A proposal for predisruption sale of call options on SPR oil was first made by Philip K. Verleger, Jr. (1981). Recently, this idea has been fleshed out and ably articulated by Alvin L. Alm (see Chapter 1).

The overall system would have many similar features to the IOCG system, except forward delivery would be optional rather than obligatory. The federal government would periodically sell call options that would be contractual obligations to sell specified quantities of oil on specified future dates and at specified "strike prices" or "exercise prices." Options could be sold at several different exercise prices simultaneously, and the amounts bid for each would indicate aggregate oil price expectations.

This would be a simpler and less complicated form of "disruption insurance" for a refiner to buy than the IOCG type of contract for forward delivery. The buyer wouldn't be obligated to take delivery and pay for the oil as of the future date but would merely be buying an inexpensive hedge as protection against future price movement. For these reasons, there would probably be a much more active market for these options than for the IOCG type of forward delivery contract.

Another distinction between the options system and the IOCG system is that the periodic sale of options would generate a steady flow of revenue for the SPR, whereas the IOCG conditional bids for forward delivery would not. Under present law, the options contracts would have to be conditioned on a presidential declaration. The system could start out on that basis and then move to a different "trigger" once the law was changed. There are many other types of triggers: (1) a decision by the Secretary of Energy, applying to only part of the SPR; (2) a decision by a special SPR board or commission appointed by the president and applying to only part of the SPR; or (3) allowing the release of part of the SPR to be automatically triggered when the spot market price reaches the option exercise price.

There are at least three kinds of misconceptions about options, all of which could be corrected by DOE educational programs and strong leadership. One confused viewpoint is that the government should not be involved with futures or options contracts because they are "risky" and "speculative." If the government sells futures or options contracts on SPR oil, it isn't taking a speculative position any more than it does when the Treasury Department holds an auction of Treasury bills. It is those who choose to buy the futures contracts or options contracts who are deciding to take a position. Options markets or futures markets don't increase or decrease the amount of price volatility risk to buyers or sellers of crude oil. All they do is redistribute that same risk from those who don't want it (e.g., an independent oil refiner) to those who are willing to take the risk for the possibility of making a speculative profit.

Second, some people feel that the sale of options with certain specified exercise prices amounts to the government "setting the price." Images are conjured up of oil price controls. This is a complete misconception. If the federal government offers options at many different exercise prices, all of them considerably higher than the current spot market price, this doesn't imply anything about what the actual price will be or should be on the future exercise date. The market will decide what value such an option would have (some options, at high exercise prices, may have no market value), and the spot market will decide what price a "wet barrel" is worth on the exercise date. The use of futures or options contracts does not involve government price setting.

A third misconception about predisruption options or futures contracts is that they would have an undesirable impact on private inventory holding. During calm years—if there is such a thing in the oil business—the presence of governmentally offered futures or options contracts on SPR oil would have a negative but negligible impact on private inventories. There would be little interest in these contracts. However, when international events begin heating up (like the 1979 Iranian situation), then the presence of these futures and options contracts on SPR oil would have a negative and substantial impact on private inventory building, and that impact is just what the SPR is supposed to accomplish. It was private inventory building in 1979 that turned a minor supply shortfall into a disastrous oil price runup. SPR futures or options would discourage inventory building just when that would be very socially desirable.

ARRAYING THE POLICY ALTERNATIVES

Table 9-1 arrays the many SPR institutional alternatives. The columns are labeled F for alternative financing mechanisms, O for alternative organizational and management configurations, P for alternative predisruption mechanisms related to later drawdown, and D for alternative disruption drawdown mechanisms. The policy decision-maker can choose one alternative from each of the four columns.

The Spectrum of Alternatives for Financing the SPR

All of the financing alternatives except the present system involve levying a user fee on crude oil input to U.S. refineries. To achieve neutrality in the cost and price impacts of such a user fee, it might also be necessary to levy an equivalent fee on imports of refined petroleum products. The user fee would be used to cover the interest payments on SPR debt plus the annual operating costs of the SPR.

If all of the SPR debt and its operating cost were to be covered by this user fee, the fee would initially be about 42 cents per barrel of refinery crude input—or about one cent per gallon of refined product output. Over the decade of the 1980s, the fee would go up about 5 cents per barrel of crude input per year (about 0.1 cent per gallon of refinery output) as new facilities and oil are added to the SPR.

It is very reasonable to ask the refiners and consumers of petroleum products to reduce the burden of SPR expenses to general taxpayers, because it is the refiners and consumers of petroleum products who are the principal beneficiaries of "SPR disruption insurance." These kind of fees are a central feature of the SPR systems in the Federal Republic of Germany, Switzerland, and the Netherlands.

Table 9-1. A Depiction of SPR Institutional Alternatives.

Financing Alternatives	Organizational Alternatives	Predisruption Alternatives	Disruption Alternatives
F–1 Present system of off-budget federal borrowing through special Treasury account	O–1 Present SPR program	P–1 Present system of no commitment of the SPR prior to a presidential declaration	D–1 Present system of spot market auctions for immediate delivery or within a few weeks; take or full pay obligation; any price bid
F–2 Same as F–1, but government covers annual debt service and operating cost from a "user fee" on refinery crude input	O–2 One federal organization but with some facilities dedicated to multiple cycle drawdown (MCD)	P–2 Sale of futures contracts conditioned on presidential declaration, minimum price, small part of SPR, full or partial penalty for nondelivery	D–2 Auction for forward delivery; take or pay full penalty; any price bid; similar to futures contract
F–3 Quasi-public corp. financed by federally guaranteed bonds and user fee on refinery crude input	O–3 Two organizations (one federal, one quasi-public): one for severe and rare disruptions; one with MCD to deal with less severe market disturbances	P–3 Sale of options contracts conditioned on presidential declaration; at various exercise prices considerably above current spot prices	D–3 Auction for forward delivery; take or pay partial penalty; any price bid; a liberal futures contract
F–4 Quasi-public corp. with "oil collateral bonds" equal to 50% of value of SPR oil, plus other federally guaranteed bonds, plus user fee	O–4 One federal organization that builds and fills facilities, and then "sells" them to a quasi-public organization that operates them	P–4 Same as P–3, but part of the SPR could be triggered by subpresidential authority—DOE secretary or a board	D–4 Auction for forward delivery; take or pay no penalty; any price; tantamount to an options contract
F–5 Same as F–4, but also use oil-price-linked bonds to the extent they are accepted by capital markets	O–5 Quasi-public organization for both construction and operation of the whole SPR	P–5 Same as P–3, but part of SPR would be automatically triggered when spot price reaches the options contract	D–5 Auction of options contracts at specified prices; not substantively different from D–4

Even more important than deciding how to meet the annual operating costs of the SPR is how to capitalize its facilities and oil purchases. Table 9-2 shows a typology of SPR bond-financing instruments. Note that all categories of bonds in Table 9-2 are interest bearing. The financing alternatives shown as F-1, F-2, and F-3 in Table 9-1 all fall in the upper left-hand corner of Table 9-2— federally guaranteed bonds that are not linked to the price of oil. While these securities are very acceptable in capital markets, whether issued by the federal government or a quasi-public corporation, they are really just additions to federal debt. They would compete with conventional Treasury debt in the governmental debt part of the capital markets.

In order to get away from the criticism that SPR debt is just a "federal deficit by another name," it is necessary to create an instrument that competes with corporate bonds rather than government bonds. As shown by financing alternative F-4 in Table 9-1, and by the upper right-hand box of Table 9-2, this could be accomplished by having the bonds collateralized by the oil in the SPR (or the proceeds from the sale of the oil). Given the volatility in the price of oil it might be necessary to issue an amount of bonds with par value equal to only, say, half the current value of the oil that served as collateral. Thus, the part of the SPR financed by oil-collateralized bonds would also have to have supplementary financing via federally guaranteed bonds.

Oil-price-linked securities may also be a supplementary method of financing the SPR, but their attractiveness in capital markets is uncertain at the present time. Crude oil futures began trading on the Chicago Board of Trade (CBT) and the New York Mercantile Exchange (NYMEX) on March 30, 1983. If, as expected, they soon become the most important kind of commodity futures contract, and if futures trading is followed by options trading *in commodity markets*, then perhaps some of that acceptability will eventually spill over into bond markets. The oil-price-linkage feature for SPR bonds could be like the convertible (to equity shares) feature in many corporate bonds. As shown in the bottom two boxes of Table 9-2, the oil-price-linkage feature can be combined with either federally guaranteed bonds or oil-collateralized bonds.

Although it is not mentioned in any of the financing alternatives of Table 9-2, the sale of call options on the SPR, both before and during disruptions, could be a significant source of revenue that

Table 9-2. A Typology of SPR Bond Financing Instruments
(*all categories are interest bearing*).

	Federally guaranteed bonds	Not federally guaranteed bonds, but collateralized by the oil assets of the SPR
Principal not linked to the price of oil	Not effectively different from the present system of borrowing through special SPR Treasury account	Oil collateralized bonds
	(F-1, F-2, F-3) ——————➤ (F-4)	
	(F-5) ◄	(F-5)
Option of adjusting the principal at maturity by an oil price index	Federal bonds, with an oil convertibility feature	Combine oil collateral and oil convertibility feature

could be combined with any of the five financing alternatives. This will be discussed below.

The Spectrum of Organizational Alternatives

Apart from whether all or part of the SPR should be financed outside of the federal budget, there are questions about whether the SPR should be divided into separate entities with separate physical facilities and management. For example, if it were decided to change the law so that part of the SPR could be drawn down prior to a presidential declaration, then this portion of the reserve would probably experience more numerous drawdowns and refills over a long period. Those entrusted with making the early release decision (e.g., the Secretary of Energy or a special board) would need to have more intimate knowledge of the day-to-day workings of the oil spot market. This form of organization (0-2) would not imply drastic differences in the types of people, management, and facilities, but there would need to be some changes from the current mode of operation of the SPR program.

Another possibility (0-3) would be to set up a quasi-public corporation just to manage and finance the multiple cycle drawdown (MCD) facilities. This organization could start out in a modest way by raising financing sufficient to "buy" the Weeks Island Facility, and perhaps a part of the Saint James terminal, from DOE at the replacement cost of the facilities and oil. This sale would give DOE something like $1.8 to 2.3 billion in off-budget capital that it needs to proceed with Big Hill and continue its oil-acquisition program. The board of directors of the quasi-public corporation owning Weeks Island could then become the decision body for drawdowns preceding a presidential declaration.

Still another concept (0-4) would be for the federal government to specialize in the construction and filling of SPR facilities and for the quasi-public corporation to specialize in operating, drawing down, and refilling. This would imply that the whole SPR would eventually be in the hands of the quasi-public corporation. There might be some objections to this alternative from the military and foreign policy parts of the federal government. Even in the Federal Republic of Germany, where a quasi-public corporation manages a 100-million-barrel reserve, a separate reserve of about 25 million barrels is operated by the government and is associated with the national security aspects of more severe international crises.

The Spectrum of Predisruption Alternatives

Present law would permit use of predisruption alternatives P-1, P-2, or P-3. All of these alternatives require waiting for a presidential declaration before any part of the SPR could be drawn down. Under either P-2 or P-3, a part of the SPR would be committed for drawdown but still contingent upon a presidential declaration.

Comparing P-2 (futures contracts) and P-3 (options contracts), the latter would probably be preferable because it would be a less complicated or risky way for refiners to hedge their cost positions as a disruption possibility grew more serious. There would be a larger market for options contracts than for futures contracts. Also, and importantly, the sale of options contracts would provide a stream of revenue for the SPR (or a quasi-public corporation handling "early phase drawdown"), whereas sale of futures contracts would not provide revenue.

Alternatives P-4 or P-5 involve "early phase drawdown" before a presidential declaration, at least for part of the reserve. This is a preferable system for the reasons described above. Under P-4, the board could be the board of the quasi-public corporation if P-4 were combined with those other alternatives for financing or organization.

Alternatives P-4 and P-5 are not mutually exclusive. When the secretary of Energy or the board of directors of a quasi-public corporation offered a given block of call options for sale, they could declare that those particular options would be "automatically triggered" once the spot price reached the option exercise price.

Some people have expressed concern that such automatic triggering could cause "false triggering." If options were periodically sold for various exercise prices, and if the prices for an option three or six months in the future were set well above the current spot price, then this would be unlikely, and it would only affect a small quantity of oil. It might even be desirable for options to be exercised once in a while so that the operation of the early phase reserve would take on a routine character and not change much as a real disruption approached. This would build confidence in the private sector that the early phase reserve was capable of operating predictably when it was really needed.

The Spectrum of Disruption Alternatives

For the SPR to perform its function during a disruption it is desirable for it to offer many alternatives to refiners. Some would want their oil immediately, some would want to be assured of delivery a few weeks or months away, and others may want only options contracts to hedge their cost position. To satisfy these diverse needs, it is probably desirable to offer all of the alternatives shown in Table 9-1. Auction of options contracts during a disruption does offer the unique advantage of earning SPR revenue. During a disruption, this revenue could be substantial.

INTEGRATION OF DRAWDOWN WITH PRIVATE FUTURES AND OPTIONS MARKETS

Several commodity exchanges have been trading futures contracts in #2 heating oil, #6 residual oil, and leaded and unleaded gasoline for several years. Crude oil futures contracts were introduced on both the CBT and the NYMEX on March 30, 1983. These contracts will also be traded on the Chicago Mercantile Exchange (CME) and the World Energy Exchange (a new computerized exchange in Dallas) later in 1983. Crude oil futures and gas oil futures are already trading on the International Petroleum Exchange (IPE) in London.

There isn't any doubt that crude oil futures contracts will succeed. Even the major vertically integrated oil companies, which initially shunned the futures markets in petroleum products, are now active participants. A futures market need not handle very much of the overall volume in a commodity to exert a strong influence on the prices of the commodity. For example, futures markets in copper only account for about 5 percent of overall trade in copper, yet they have a strong influence on copper pricing. Likewise, the number of transactions in futures that involve an actual delivery of the commodity are expected to be a very small percentage of total futures transactions. For the petroleum product futures markets, this proportion started out at about 3 to 4 percent and then decreased to 1 to 2 percent as the trading volume grew.

The relevant question is not whether futures trading in crude oil will succeed but which of the exchanges will dominate. Even though there are many commodity exchanges throughout the world, there is a strong tendency for one or two exchanges to dominate the others in futures trading of a given commodity. This is because a larger trading volume enhances the liquidity of each contract, which in turn attracts more trading—a classic economy of scale phenomenon.

Relevance of Futures Contracts to the SPR and an Early Phase Reserve

Most of the SPR sites are salt domes. They are designed for five or less cycles of drawdown and refill. They are limited in the number of cycles by the fact that water injection into the salt dome to accomplish oil drawdown also results in leaching of salt and enlargement of the volume of the salt dome by 15 percent during each drawdown cycle. The enlargement can cause structural weakness if it proceeds too far. This problem might potentially be mitigated or eliminated at one or more sites by injecting salt brine instead of water. At present, there are no storage facilities for salt brine at the sites. Because the Weeks Island facility is a conventional mine rather than a salt dome, it is drawn down by sump pumps instead of injecting any liquid. This is what makes Weeks Island the logical choice for the first facility with multiple cycle drawdown capability—a necessary attribute for an Early Phase Reserve. As shown in Table 9-3, the oil stored at Weeks Island is all sour crude; that is, it has a sulphur content greater than 0.5 percent.

The question of which exchange wins the race to dominate crude oil futures, and thus which exchanges have to drop out, is relevant to the drawdown of the SPR in general, particularly the drawdown of an Early Phase Reserve. As indicated in Table 9-3, the CBT and the World Energy Exchange both use the St. James Louisiana terminal (entry point for the Capline pipeline) as the geographical reference point in their futures contracts. Both the Weeks Island and Bayou Choctaw SPR sites are connected to St. James, and the SPR also has extensive terminal facilities at St. James. The NYMEX and the CME both use Cushing, Oklahoma as the geographical reference point for their crude oil futures contracts. Cushing is the intersection point of several pipelines, including the Seaway pipeline (connected to the Bryan Mound SPR site) and the Texoma pipeline (connected to SPR sites at West Hackberry and Sulphur Mines, and the least developed SPR site at Big Hill).

As indicated by Table 9-3 and the above discussion, there is a mismatch between types of crude stored at the SPR sites and the types of crude specified in the private futures contracts keyed to the same pipelines. This would not prevent futures contracts from playing a useful role during an SPR drawdown. To the extent that the price movements of different crudes were correlated during a disruption, holding a hedge in the form of any crude oil futures contract would provide financial protection for users of other crude types. A user need not hold a futures contract corresponding to exactly the crude type he uses. However, some less sophisticated crude users might feel more comfortable utilizing futures contracts if there were a closer match between the crude types and locations in the futures contracts and their own requirements.

The mismatch shown in Table 9-3 could inhibit some other potential forms of integration between the SPR and private futures markets. If there were a closer match, then it might be possible to use part of the SPR as the "supplier of last resort" to make sure that delivery provisions of futures contracts are satisfied during a disruption.

Table 9–3.   Lack of Integration between SPR Sites and Private Futures Contracts.

	Sweet Crude	Sour Crude
St. James, Louisiana reference point	CBT contract WEX contract	
	Bayou Choctaw 18.4 million barrels	Weeks Island site 72.6 million barrels Bayou Choctaw 26.1 million barrels
Cushing, Oklahoma reference point	NYMEX contract CME contract	NYMEX contract
	Bryan Mound 64.4 million barrels West Hackberry 21.1 million barrels	Bryan Mound 39.8 million barrels West Hackberry 35.8 million barrels Sulphur Mines 13.2 million barrels

Source: Commodity exchange staffs; and U.S. Department of Energy (1983: 7).

Also, if there were a closer match, the government could hold sales for future delivery during a disruption, and these "warehouse receipts" could be the basis for securing private futures contracts. Obviously, the mismatch problem can be mitigated either by changes in the specifications in private futures contracts or changes in the geographical mix of crude types in the SPR.

Relevance of Options Trading

Officials at the CBT and the NYMEX both expect crude oil options trading to be introduced on their exchanges in early 1984. This development will facilitate the use of options by an Early Phase Reserve or the regular SPR. There would be no need for the federal government to wait until crude oil options were well established in the private sector before using them for an Early Phase Reserve. A call option on the Early Phase Reserve would be a relatively simple instrument with great acceptability and liquidity, since there would be no doubt that the oil existed and could be delivered if the option were exercised.

If the federal government does issue its own futures contracts or options contracts, there ought to be no restrictions on the transferability of those contracts. If they trade on commodity exchanges, they will become more attractive instruments, they will raise more revenue for the SPR, and they will be more successful at reducing economic losses during disruptions.

CONCLUSION

Among the problems and challenges facing the SPR program in 1983, three are most urgent:

1. The continued acquisition of SPR oil and the development of the Big Hill site beyond fiscal year 1984 may be in danger because of the overall federal budgetary pressures. It may be useful to consider alternative methods of financing, at least for part of the SPR.
2. Under EPCA, no part of the SPR can be drawn down without a presidential declaration that a "severe energy supply interruption" exists. Because a decision of that kind would likely be de-

layed in order to avoid alarming allies and foreign oil suppliers, a lot of oil price runup and economic damage would likely occur before the SPR was drawn down. Mechanisms need to be developed to draw down part of the SPR at an earlier date via subpresidential authority.

3. The SPR emergency drawdown plan announced by DOE in December 1982 contemplates using only spot market auctions of SPR oil, and only after a presidential declaration. Other market-oriented approaches to drawdown may be able to influence spot market prices at an earlier phase of a disruption.

NOTES TO CHAPTER 9

1. See Plummer (1982a).
2. In the area of military preparedness, the Defense Department has five levels corresponding to different degrees of seriousness, going from DEFCON #5 (defense condition #5) of demobilization up to DEFCON #1 of a declared war. Up through DEFCON #3 (alert), the decision is made by the Joint Chiefs of Staff rather than the president. The problem with EPCA is that there are no intermediate subpresidential degrees of deployment of the SPR.
3. For a discussion of the pros and cons of the DOE emergency drawdown plan, see Verleger (1983) and Treat (1983).
4. See Mehle (1981), Plummer (1981), Safer (1981), and Arditti (1981).
5. This group consists of Armco Corporation, Bethlehem Steel, Cone Mills, Firestone Tire and Rubber, General Motors, Kimberly Clark, and U.S. Steel. Its proposals are contained in a November 5, 1982 letter to DOE commenting on DOE's plan for drawing down the SPR via spot market auctions and also in an IOCG statement of February 17, 1983 to the House Subcommittee on Fossil and Synthetic Fuels.
6. See Chapter 10. This research indicates that futures contracts dampen inventory building and thus produce a lower price trajectory during all stages of a disruption.

REFERENCES

Alm, A.L., and E.N. Krapels. 1982. "Building Buffer Stocks in a Bear Market: Policy Choices for Emergency Oil Reserves." Harvard Energy Security Program Discussion Paper Series, H-82-01.

Arditti, F.D. 1981. Testimony before the Senate Subcommittee on Energy and Minerals, Washington, D.C., June 12.

Mehle, R.W. 1981. Testimony before the Senate Subcommittee on Energy and Minerals, Washington, D.C., June 12.

Plummer, J.L. 1981. Testimony before the Senate Subcommittee on Energy and Minerals, Washington, D.C., June 12.

Plummer, J.L., ed. 1982a. *Energy Vulnerability.* Cambridge, Mass.: Ballinger Publishing Company.

Plummer, J.L. 1982b. "Why the U.S. Should Manage the SPR as a Private-Public Corporation." *Energy Management* (November/December).

Safer, A.E. 1981. Testimony before the Senate Subcommittee on Energy and Minerals, Washington, D.C., June 12.

Treat, J.E. 1983. Testimony before the House Subcommittee on Fossil and Synthetic Fuels, Washington, D.C., February 17.

U.S. Department of Energy. 1983. "Strategic Petroleum Reserve Annual Report 1983." Washington, D.C.: U.S. Government Printing Office.

Verleger, P.K. 1981. "Let the Market Fill the U.S. Petroleum Reserve." *Wall Street Journal*, April 29: editorial page.

Verleger, P.K. 1983. Testimony before the House Subcommittee on Fossil and Synthetic Fuels, Washington, D.C., February 17.

10 DRAWING DOWN THE STRATEGIC PETROLEUM RESERVE
The Case for Selling Futures Contracts

Shantayanan Devarajan and R. Glenn Hubbard

With over 300 million barrels of oil in place, the U.S. Strategic Petroleum Reserve (SPR) represents a potentially significant tool for reducing the economic damage attendant to oil supply disruptions. However, the ability of the SPR to blunt the oil price increases during disruptions depends critically on how it is drawn down. In this chapter, we propose a particular drawdown strategy for the SPR—one that addresses an often overlooked issue, namely, the response of private stockpiles to an oil supply interruption. While several studies have pointed out that rapid and discontinuous changes in the price of oil can create problems of a special nature (Eckstein 1979; Hubbard and Fry 1982; Mork and Hall 1980), only a few have observed that the price increase caused by the supply disruption is typically heightened by the role of private crude oil inventories (Danielsen and Selby 1980; Verleger 1982). Driven by uncertainty over future supplies and the expectations of profits from still higher prices in the future, private inventory accumulation at the onset of the disruption augments world oil demand, putting further upward pressure on oil prices.

To the extent that private inventory accumulation has contributed to the great price increases during disruptions, it is desirable to develop policy instruments that could manipulate private sector

We thank Richard Ericson, James Hamilton, William Hogan, Albert Nichols, Richard Norgaard and Robert Weiner for helpful comments.

stockpiling. In this chapter we analyze one such instrument: the sale of futures contracts in SPR oil. We show how the existence of SPR futures contracts dampens private inventory demand by guaranteeing future supplies. In addition, futures market sales are compared with direct, spot market sales. We conclude that the relative benefits of a futures over a spot market sale from the SPR is an empirical question. We attempt to shed some light on that question by simulating a disruption in a model of the world oil market and measuring the effects of futures and spot market sales of SPR oil. Finally, we summarize the salient points and consider the implications of our results for policy.

FUTURES VERSUS SPOT MARKET SALES OF THE SPR

To make the case for futures market sales of SPR oil, it is important to examine first the rationale behind the private inventory behavior during a disruption. Economic optimization suggests that the speculative component of stocks be built when expected future prices are high relative to today's price, and that they be drawn down when expected future prices are low. A characteristic of disrupted oil markets is the continued increase, rather than a one-shot jump, in the spot market price. There are several reasons for this. The size of the shortfall may increase over time; there is much confusion initially about the magnitude, duration, and even existence of a disruption; since most oil is traded in long-term contracts, agents enter the spot market at different points in time. Whatever the reason, the fact that rising future prices can be identified with disrupted markets makes the observed accumulation of private inventories hardly surprising.

However, because of imperfections in the domestic economy as well problems with monetary and fiscal policy responses, every barrel of imported oil imposes a cost on the United States over and above the private cost to the importer. Indeed, this is the motivation behind the proposal for an import tariff during interruptions (Hogan 1982; Verleger 1982). The disruption-induced increase in private inventories, therefore, contributes to this social cost insofar as it is a component of import demand. When one considers that the spot market price—which is bid up by all this "panic buying"—acts as a trigger for various crisis-level activities, the case for government inter-action becomes all the more compelling.[1]

The link between expected future prices and speculative hoarding provides an opportunity for policy intervention if the government can manipulate price expectations. Here the SPR can play a role. The announcement of an SPR drawdown strategy for the future means more oil will be available at the future date, which lowers the ex-pected future price and, therefore, expected future profits from speculation. This, in turn, reduces inventory demand and spot prices today. Of course, the success of this intervention depends on how credible the announced SPR drawdown strategy is. If firms suspect that the government may not release the oil in the future, they will continue to hold speculative stocks. By selling futures contracts, however, the government can guarantee that a specified amount of SPR oil will be available at a certain time in the future. Thus, instead of its traditional risk-sharing role, the futures contract here plays the role of a credible guarantor of the government's SPR drawdown policy.[2]

By selling futures contracts and thereby agreeing to deliver SPR oil at some future date, the government has effectively decreased world oil demand at that date, putting downward pressure on both the expected future spot price (by directly reducing future demand) and today's spot price (by reducing speculative demand). The govern-ment has thus been able to depress the spot market price of oil today *without releasing a drop of oil from the strategic reserve.*

The government can also lower the spot market price by selling SPR oil directly in the spot market. However, because of differences in the way they affect expectations about the future, spot sales and futures sales from the SPR will result in different sequences of oil market prices.

To see this, consider a model where the spot market price of oil is determined by equating supply of oil with demand for oil in the spot market. The latter includes a component for speculative inventory demand which, as we said earlier, depends on the expected future price of oil. Thus, expectations of future prices affect the spot mar-ket price of oil via their impact on demand for inventories.

How are expectations of future prices formed? It is reasonable to expect that they are formed by buyers' and sellers' expectations of supply and demand conditions in the future. We will see that the difference between spot and futures market sales of the SPR have to do with their effects on supply conditions in the future.

When the government sells, say, a million barrels in SPR futures contracts for delivery three months from the date of sale, it is guaranteeing that the supply of oil three months hence will be a million barrels greater than it would have been without the sale. Consequently, buyers will lower their expected future price, which in turn will dampen inventory demand today, lowering today's spot market price.

Of course, if the government sold a million barrels in the spot market, that too would lower the spot market price.[3] The relative benefits of the two sales strategies depend on how the combined effects of falling expected future prices from increases in future supply and the response of inventory demand to a drop in the expected future price stack up against the response of spot prices to increases in spot supply.

So far, we have contrasted spot and futures market sales in terms of their effects on the current spot price. What about their effects on the realized spot price in the future—say, for example, the spot price three months from now? Here, the results are unambiguous. If a million barrels of futures contracts were sold today, they would lower effective demand (or increase effective supply) by a million barrels in the spot market three months from now. If the same amount of oil were sold in today's spot market, it would lower effective demand three months from now only to the extent that any of this oil is resold when that market reopens: The *most* by which demand can be reduced is a million barrels. Thus, the spot price in the future under a futures sale is a lower bound on the spot price under spot sales from the reserve.[4]

At the very least, therefore, the trajectory of spot prices will be different between spot and futures sales from the SPR. The question we will turn to in the next section is "How different?"

SIMULATION RESULTS

In this section we examine the impact of a futures sale of SPR oil on private inventory behavior and compare the effectiveness in reducing oil prices of spot and futures market sales. We use the model of the world oil market and the U.S. economy described in detail in Hubbard and Fry (1982). The model allows changes in world oil prices to affect U.S. economic performance and vice versa. The details of the model's specification are given in Appendix A.

We first simulate an undisrupted market from the first quarter of 1982 (1982:1) to the fourth quarter of 1985 (1985:4). The disruption that we simulate is a reduction in OPEC capacity of 6.5 million barrels per day (mbd) for the whole of 1983; the capacity is restored in 1984. The resulting price increase is not as large as one would expect from the spot price formulation alone because of increased production from nondisrupted OPEC members, the drop in consuming country demand (because of the higher prices), and the soft market at the onset of the disruption. Table 10-1 compares the paths of crude oil spot prices and U.S. private inventory accumulation (of crude oil and petroleum products) across the two scenarios. The assumed accumulation pattern for the U.S. SPR is also given.

Two results illustrated in Table 10-1 are particularly interesting. First, spot prices rise fairly sharply after the onset of the disruption

Table 10-1. Comparison of Control Scenario and Disruption Scenario.

	Spot Price ($/barrel)		Private Stock Change (1,000 barrels/day)		SPR Fill (1,000 barrels/day)	
	C	D	C	D	C	D
1982: 1	31.70	31.70	−640	−640	200	200
2	31.60	31.60	700	700	200	200
3	31.20	31.20	490	490	200	200
4	31.20	31.20	−670	−670	200	200
1983: 1	30.60	36.80	−1080	−910	200	0
2	29.60	45.30	390	700	200	0
3	28.30	54.70	570	960	200	0
4	27.10	59.50	−470	−130	200	0
1984: 1	26.20	51.90	−1250	−1180	200	200
2	26.00	45.10	190	370	200	200
3	26.90	39.40	480	580	200	200
4	28.50	35.20	−360	−480	200	200
1985: 1	30.40	35.10	−1070	−1250	200	200
2	32.10	31.50	270	80	200	200
3	34.50	29.20	560	360	200	200
4	37.40	28.00	−230	−520	200	200

in 1983: 1. (By the end of the solution interval, spot prices are actually lower in the disrupted scenario because of the large reductions in oil demand. The marginal cost of a barrel of oil to U.S. refineries is still slightly higher by 1984: 4 in the disruption scenario.) Second, private inventory behavior responds quickly to expected future profits. Inventory accumulation in 1983: 2 to 1983: 4 in the disruption scenario ran almost 100 percent higher than in the base case. Stocks are decumulated relative to the control by the end of the interval, so that an inventory level equilibrium is restored.

We can now test the effectiveness of two SPR drawdown policies: (1) a spot market sale of 1 mbd during 1983: 2 and (2) a sale of futures contracts in 1983: 2 to sell 1 mbd in 1983: 3. To refill the depleted 90 million barrels, we assume in both cases that the SPR fill rate after the disruption (i.e., in 1984 and 1985) will be 300,000

Table 10-2. Comparison of Spot Market and Futures Market Oil Sales.

| | Spot Price ($/barrel) | | | U.S. Private Stock Change (1,000 barrels/day) | | |
Quarter	Base Disruption	Spot Sale	Futures Sales	Base Disruption	Spot Sale	Futures Sales
1982: 1	31.70			-640		
2	31.60			700		
3	31.20			490		
4	31.20			-670		
1983: 1	36.80	36.80	36.80	-910	-910	-910
2	45.30	42.50	44.05	700	570	480
3	54.70	52.70	49.20	960	960	850
4	59.50	58.00	55.20	-130	-120	-110
1984: 1	51.90	51.10	49.30	-1180	-1180	-1170
2	45.10	44.70	43.30	370	370	350
3	39.40	39.00	38.10	580	570	570
4	35.20	35.40	34.60	-480	-460	-480
1985: 1	35.10	35.80	35.40	-1250	-1230	-1210
2	31.50	31.60	31.50	80	100	100
3	29.20	29.70	28.80	360	380	380
4	28.00	28.85	27.70	-520	-570	-490

barrels per day instead of 200,000 barrels per day. Table 10-2 compares the effect on spot prices and on U.S. private inventory accumulation for the two strategies.

Looking at the results in Table 10-2, we see that both strategies influence spot prices and private inventory behavior. The size of the impacts is dependent upon the fact that policy changes are examined in a "U.S. only" analysis; that is, foreign inventory accumulation assumptions are held constant across scenarios (see Chapter 11 for a discussion of international interaction).

Two interesting patterns surface. First, while the direct drawdown (spot market sale) of the SPR oil in 1983: 2 lowered the spot price in that quarter, the sale of futures contracts in SPR oil for delivery in 1983: 3 also lowered the spot price immediately. Moreover, under the futures sale option, spot prices remained slightly lower for the rest of the interval.

Second, the drawdown strategies blunted some of the extra private stockpiling that occurred because of the higher expected prices during the disruption. Comparing the private stockpiling trajectories in Table 10-2 reveals that part of the spot market sale was added to private stocks. By guaranteeing supplies to the market for 1983: 3, the futures sale reduced private inventory accumulation during the disruption by more than the spot market sale. By 1985, both paths returned to that prevailing in the absence of policy intervention.

CONCLUSION

In this chapter we have examined the potential benefits of drawing down the SPR by selling futures contracts. Observing that, during oil supply disruptions, private inventories act to exacerbate the sharp rise in oil prices, we showed how sales of SPR futures, unlike other policy responses, dampen inventory demand by affecting expectations of future oil prices. Using a two-period model we showed how the sale of an equivalent amount of SPR oil in the spot and in the futures market could lead to a more favorable sequence of spot prices in the latter case. Our simulation results, based on a model of the world oil market and the U.S. economy, indicate that futures sales (1) achieved much of the price-reducing benefits in the early stages of a disruption and (2) led to a lower price trajectory overall when compared with spot market sales.

In addition to resulting in a more favorable spot price trajectory, drawing down the SPR by selling futures contracts has at least four other advantages. First, the sale of SPR futures permits the government to achieve significant benefits, in terms of a reduced spot price, before releasing a drop of oil from the strategic reserve. This is important because, as many authors have stressed, delays in decision-making can be extremely costly during interruptions. To the extent that the decision to sell futures contracts is easier to make than the decision to draw down the SPR physically, the futures option "buys time" for the government (Hogan 1981: 277).

Second, even when the contracts come due, the government may be able to avoid drawing oil from the reserve. If the interruption is over before the due date, the contracts would have been priced higher than the now normal spot market price. The government can meet its contractual obligation in this case by buying oil in the spot market and giving it to the holder of the futures contract. Alternatively, the government could just buy back the futures contracts. Even if the interruption is not over when the contracts mature, the government can purchase oil in the spot market to meet its obligations. Of course, this will forego the price-reducing benefits of the futures sales described earlier. However, it is worth noting that if for reasons of national security, say, the government is reluctant to withdraw SPR oil, it need not do so under a regime of futures contracts.

Third, by selling SPR futures, the government is effectively betting on the future price of oil. If the price turns out to be lower than forecast—the disruption ends sooner than expected, say—the government has registered a revenue gain. If, on the other hand, the price in the future rises even higher than was reflected in the sale price of SPR futures, the government has lost money. However, this revenue loss is consistent with the proposed stabilization rules for energy emergencies (Mork 1981), namely, that the government should follow an expansionary policy to counteract the disruption-induced slowdown in economic activity.

Fourth, SPR sales in the spot market will end up being held as inventories, so that the oil is simply transferred from the public to private reserves. By selling futures contracts, the same ownership transfer is accomplished without having to transport any oil; the social costs of transportation are therefore saved.

To be sure, this chapter represents a first pass at analyzing a new policy option for responding to a disruption in the oil market. Both the theoretical and empirical analyses can be extended to incorporate more realistic features of the world oil market. Yet, we suspect the qualitative nature of our results will remain unchanged: selling futures contracts in SPR oil can be at least as effective in dampening spot market prices as a direct sale in the spot market.

On the other side of modeling and estimating the benefits of selling futures contracts in SPR oil there lies the question of how such a scheme may be implemented. As there currently exists only a small futures market in crude oil, and as attempts to set up such a market in the past have met with limited success, the procedure by which the government could sell SPR futures contracts deserves considerable attention. Although we believe sales of SPR futures during a disruption are feasible, and in many ways easier to effect than spot market sales, we also feel that the success of such a policy depends on the amount of thought given to its implementation before the interruption occurs.

NOTES TO CHAPTER 10

1. Our analysis (and, indeed, that of other authors) is based on the assumption that the disruption is temporary. If the disruption is permanent, then it is not clear that releasing the SPR is a prudent strategy. Hence, in this chapter we are taking as given the decision to use the SPR and are comparing the relative merits of spot and futures market sales.

2. It is reasonable to ask why futures markets in crude oil are underdeveloped. In addition to the institutional reasons given by Safer (1979: 92) and Plummer (1982: 142), we can pose possible economic reasons. During "normal" times, there is not much variability in crude oil prices. While there is a lot of variability in prices during an interruption, finding short-sellers is likely to be difficult. Supplies of oil are constrained and holders of available stocks expect rapidly rising prices. Having the SPR as a participant in a crude oil futures market would be valuable both as a marginal supplier of oil and as an insurance mechanism.

    Plummer (1982: 142) points out that "there is a fallacy of composition involved in thinking of futures markets as an overall protection against oil supply disruptions." While it is true that such an option should be only part of a complete energy policy, Plummer's contention that "aggregate risk has not been reduced" (1982: 142) is not completely true. To the extent that a futures market, either in the United States alone or as an international effort, can manipulate private sector stockpiling and reduce the price runup during supply interruptions, the consuming nations' *economic* risk has been reduced.

3. A spot market sale of SPR oil may also affect future price expectations to the extent that some of the oil may be resold in the future. However, it cannot affect price expectations more than a futures sale would, since, with a spot sale, there is uncertainty over which date in the future the oil will be resold. Typically, a fraction of the oil can be expected to be resold three months in the future. By contrast, with a futures sale, agents can expect all the oil to be available at the date the contract is due. Hence, the impact on the expected future price will be greater with a futures sale than with a spot sale.

4. This conclusion holds a fortiori in a multiperiod world, so long as there is no further policy intervention in the subsequent periods.

REFERENCES

Danielsen, Albert, and Edward B. Selby, Jr. 1980. "World Oil Price Increases: Sources and Solutions." *Energy Journal* 1, no. 4 (October): 59–74.

Eckstein, Otto. 1981. "Shock Inflation, Core Inflation, and Energy Disturbances in the DRI Model." In *Energy Prices, Inflation and Economic Activity*, edited by K. Mork. Cambridge, Mass.: Ballinger Publishing Company.

Hogan, William W. 1981. "Import Management and Oil Emergencies." In *Energy and Security*, edited by D. Deese and J. Nye. Cambridge, Mass.: Ballinger Publishing Company.

_____. 1982. "Oil Stockpiling: Help Thy Neighbor." Harvard Energy Security Program Discussion Paper Series, H–82–02.

Hubbard, R. Glenn, and Robert C. Fry, Jr. 1982. "The Macroeconomic Impacts of Oil Supply Disruptions." Harvard Energy and Environmental Policy Center Discussion Paper Series, E–81–07.

Mork, Knut A. 1981. "Macroeconomic Analysis of Energy Price Shocks and Offsetting Policies: An Integrated Approach." In *Energy Prices, Inflation and Economic Activity*, edited by K. Mork. Cambridge, Mass.: Ballinger Publishing Company.

Mork, Knut A., and Robert E. Hall. 1980. "Energy Prices, Inflation and Recession." *Energy Journal* 1, no. 2 (April): 41–53.

Plummer, James L. 1982. "U.S. Stockpiling Policy." In *Energy Vulnerability*, edited by J.L. Plummer. Cambridge, Mass.: Ballinger Publishing Company.

Safer, Arnold E. 1979. *International Oil Policy*. Lexington, Mass.: D.C. Heath.

Verleger, Philip K., Jr. 1982. *Oil Markets in Turmoil*. Cambridge, Mass.: Ballinger Publishing Company.

11 GOVERNMENT STOCKPILES IN A MULTICOUNTRY WORLD
Coordination versus Competition

R. Glenn Hubbard and Robert J. Weiner

Three times in the past decade, the world has witnessed major disruptions in the supply of crude oil from the Middle East. Twice—in 1973–74 and again in 1979—the result has been havoc in the international oil markets and substantial damage to the Organization for Economic Cooperation and Development (OECD) economies. But the experience was not repeated when the Iran–Iraq War broke out in late 1980. General consensus attributes the relatively high level of stocks at the war's outset with facilitating the ensuing drawdown, thereby making up part of the loss and easing pressure on the spot markets. In contrast, world stock levels were below historical averages in the last quarter of 1978. The ensuing scramble to build up stocks is widely credited with exacerbating the price effects of the relatively small Iranian disruption. In this chapter we undertake an economic analysis of stock behavior and, utilizing an econometric model that links the world oil market and the economy, investigate the effectiveness of so-called "stock policies" in avoiding a repetition of 1979.

The Iranian crisis left in its wake numerous scholarly studies, many of which were variations on the theme "What happened, and how can we avoid similar disasters in the future?" Among the con-

Thanks are due to Albert Danielsen, Shantayanan Devarajan, James Hamilton, and William Hogan. However, the views expressed herein, as well as any errors, are solely ours.

197

clusions reached was that cooperation among importing countries is highly desirable but damnably difficult.

Several economists have constructed elegant optimizing models in order to quantify the potential benefits of cooperation. Such analyses have been carried out by Chao and Peck (1980), Hogan (1982), Manne (1982), and Rowen and Weyant (1982). The estimated value of cooperation varies due to the different modeling techniques employed as well as the diverse assumptions about OPEC behavior, the structure of oil demand, the size and length of the disruption, the form the cooperation takes, and underlying economic factors. Although these benefits are measured along different dimensions (e.g., cost of protective policies and changes in GNP) and are thus difficult to compare across studies, a consensus exists that the value of joint action is substantial.

These studies are generally silent on the difficulty of designing and implementing a joint program, although Hogan uses his model to demonstrate that the benefits to the United States of reneging are small compared to the risk of having an agreement fall apart altogether. In actuality, the International Energy Agency (IEA), an organization within the OECD, has been charged with the development and refinement of such a program.[1]

It is not our task here to provide a detailed critique of past IEA actions.[2] Suffice to say the consumer cooperation has not always been a resounding success; indeed, it has sometimes proved difficult to detect. It will, however, serve to summarize briefly the manner in which cooperation is to take place within the IEA. The salient points are three. First, member countries are required to hold stocks equal to ninety days of net imports. Second, the agency's oil-sharing mechanism is essentially dormant until such time as its members determine that a severe disruption has occurred. In order for the emergency program to be set in motion (referred to as "triggering"), the disruption must lead to a loss of at least 7 percent of IEA consumption—compared to a base period of the previous four quarters with a one-quarter lag for data collection.[3] Third, when the trigger is pulled, member countries are required to institute demand restraint to reduce consumption by 7 percent, reduce imports by more than 7 percent, and make up the difference by drawing down reserves.[4] These measures are designed to restrict demand in the short run in an effort to prevent oil prices from skyrocketing.

Various technical problems with such a program have been pointed out in the literature; here, we take note of two broader difficulties. First, the 7 percent threshold corresponds to a severe disruption. Assuming free world oil consumption of roughly 50 million barrels per day (mbd), and taking the IEA share of consumption as constant,[5] a loss of 3.5 mbd (net of increased exports by other producers) is necessary to trigger the emergency mechanism. The Iranian crisis, during which oil prices more than doubled, was of considerably lesser magnitude. Second, demand restraint proved to be easier said than done; the March 1979 agreement to reduce consumption by 5 percent was not accompanied by vigorous enforcement.

Among the lessons to come out of the 1979 and 1980 supply shocks was that while high stockpile levels are a sine qua non for the functioning of international sharing agreements, it is the drawdown (or buildup) behavior that is likely to spell the difference between containment and disaster. Another is that actions taken in a "subtrigger disruption" (one falling beneath the threshold) may serve to avert a 1979-style catastrophic price runup. Demand restraint having failed, the economic damage attending a subtrigger disruption has called forth proposals for coordinated drawdown programs.[6] These proposals, often termed "flexible stock policies" have generated only moderate interest in the United States. The Reagan administration officially supports IEA cooperation in the case of a large (trigger) disruption but prefers to rely on the market for anything smaller.[7] We are unaware of any economic analysis of the effectiveness of these programs in mitigating disruption damage; such is our task in this chapter.

The Strategic Petroleum Reserve (SPR) figures prominently in the U.S. government's energy emergency policy, and its fill has been expedited.[8] How and when to use the reserve is currently under study; it is by no means a foregone conclusion that the government considers a repeat of the 1979 debacle to be an acceptable outcome of market processes, or that SPR releases would be held off until mandated by international obligations.[9] Criticism of the IEA trigger often overlooks the fact that some type of threshold is necessary to any contingency plan. The threshold value is a more difficult question and constitutes part of the debate.

In evaluating SPR drawdown strategies, international coordination considerations must be kept in mind. How effective will SPR draw be in relieving pressure in the world oil market if other IEA members do not do likewise? Or, even worse, what if some countries fill while others draw? The existence of this problem is well known but, notwithstanding a few relatively superficial observations, analysis of the magnitudes involved is lacking. It is to this task that we now turn.

THE WORLD OIL MARKET

Be it explicit or implicit, a model of the world oil market is crucial to any discussion of stocks policy. Essential to our analysis is a rigorous notion of market "tightness"; we thus abstract from much of the rich institutional detail of the oil trade in order to focus on this aspect. We employ two prices as proxies for the many prevailing in the market at any given time. Crude oil is sold under term contracts at the contract price. The spot price is paid for oil purchased on a single-cargo basis.[10]

The contract price is set by OPEC in accord with its production decisions and demand estimates.[11] Given the difficulty in forecasting demand and the numerous minor shocks inherent in any market, the contract price will not always equate supply and demand. The spot market serves to satisfy the excess and thus acts as a signal of market disequilibrium to OPEC, which adjusts the contract price. The process is then repeated.

The spot price increases when the market tightens. Two forms of tightening are possible: Demand can increase due to changes in consumption or stock buildup, and supply can decrease due to disruption in a producing country or deliberate production cuts. In order to capture both effects, we employ a form of price reaction function (see Figure 11-1), following Nordhaus (1980), Plummer (1981), and Hubbard and Fry (1982).

When a disruption occurs, capacity is removed from the market, and the output-to-capacity ratio of the nondisrupted producers rises. At higher prices these producers are willing to accelerate output, thereby bumping up against their own capacity constraints. When excess capacity no longer exists, (output/capacity = 1), even large increases in the spot price can elicit little further supply response; hence the nonlinearity of the curve.

Capacity decisions are assumed to be determined by longer term considerations outside the scope of the model and are taken as exog-

Figure 11-1. OPEC Price Reaction Function.

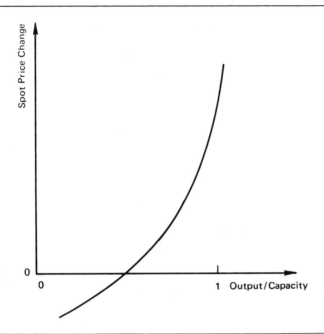

enous. OPEC output is calculated from the condition that supply and demand be equal, where demand comprises U.S. and foreign consumption, SPR fill, U.S. private inventory accumulation, and foreign inventory accumulation. Supply is divided according to source: non-OPEC, disrupted OPEC, and nondisrupted OPEC.

U.S. consumption is assumed to depend on the domestic refiner's acquisition cost, income, and a vector of structural variables, including past prices. Foreign consumption is defined similarly.

The U.S. refiner's acquisition cost is taken to be an average of spot and contract prices (plus transport costs), and to adjust to the spot price. The domestic price abroad is defined similarly.

The objective of stocks policy should be clear: to reduce demand for OPEC output, thus moderating increases in the output-to-capacity ratio and reducing pressure on the spot price. Stock policy yields "more bang for the buck" as a disruption worsens due to the nonlinearity of the price-reaction function.

We are interested in the effect of SPR releases on the spot price. In this model, such effects are three. The "direct effect" is to ease pressure as the SPR release reduces demand for OPEC output. The "feedback effect" occurs because holding down the spot price serves to hold down domestic prices at home and abroad as well, thus reducing the cutbacks in U.S. and foreign consumption. The feedback effect clearly works against the direct effect. Together, these two constitute what we shall call the "domestic effect." The "international interaction effect" depends on the reaction of foreign stock to SPR releases. "Cooperation" implies that the SPR and foreign stockpiles are drawn down simultaneously. Under "competition" foreign stocks may be built up as the U.S. government draws down the SPR. A mathematical expression for these effects can be found in Appendix A.

Cooperation serves to magnify the benefits of the SPR release; competition serves to mitigate them. In the extreme case, when foreign stocks are built barrel-for-barrel as the SPR is released, the net effect on the oil market is nil. It is important to realize that due to the nonlinearity of the price reaction function, the magnification (or mitigation) effect is more than proportional. Thus, as in the analyses referenced above, cooperation provides more than proportional benefits.

Finally, we should note that the benefits of stock draw are likely to be underestimated by the above analysis, which covers only "within-drawdown period" effects. Insofar as current spot prices influence future spot prices, the effect of an SPR drawdown today will be felt in future periods as well.

The Role of Private Stocks

For the sake of exposition, attention thus far has focused on strategic interaction between the SPR and foreign stocks. The SPR, however, constitutes but a small fraction of U.S. inventories. The bulk of the rest is controlled by the petroleum industry, which is free to build or draw as it sees fit. Indeed, folklore has it that private companies were largely responsible for destabilizing the oil market in 1979 through stock building.

Some have suggested that this stocking up represented a panic reaction by those seeking supplies "regardless of price." A related, more sophisticated line of reasoning holds that firms build stocks as prices rise, only to draw them down when the market slackens. The view of stock change as procyclical implies that inventory behavior necessarily serves to exacerbate the problems associated with disrupted markets (Frankel 1982; Danielsen and Selby 1980). Drawing down the SPR makes little sense in this type of world.

Verleger (1982) has advanced an alternative theory—one that centers on the relationship between stock changes and potential profits. Verleger predicts stock build when the difference between expected future prices and current prices exceeds the cost of storage. As long as prices are expected to rise sufficiently, the predictions of the two theories are identical. These expectations cannot continue indefinitely, however; eventually, stocks act as a stabilizing force.

The empirical investigation presented here takes as inspiration the work of Verleger. As he points out (1982: 120), "[i]t is difficult to model the behavior of oil industry inventories." His examination of the profit motive produced weak (i.e., correctly signed but largely statistically insignificant) results, implying that the "panic theory" could not be rejected. The specification estimated here contains a profit motive but differs from the one employed by Verleger in that: (1) the stock and consumption data have been seasonally adjusted (to account for the normal patterns of alternating build in the second and third quarters and draw in the fourth and first quarters); (2) effects of unanticipated changes in demand ("surprises") are incorporated; (3) we allow for the effects of price controls on stock behavior; and (4) we do not lump the disparate countries of the OECD together.

We take the inventory-to-sales ratio, rather than the raw inventory level, as the variable to be explained. Although not strictly defensible on the grounds of economic theory alone, the use of this ratio is consistent with trade and industry characterization of stocks in terms of "days of consumption."

Our explanatory variables are three. The first attempts to capture the profit motive by comparing expected future prices with today's prices and carrying costs. We expect stock build when the difference is positive. The second reflects the cost of stock adjustment by comparing the inventory-to-sales ratio in the last quarter with a four-quarter moving average. If last quarter's value is high relative to trend, we expect a gradual movement back. Finally, the "demand surprise" term is included because higher than expected demand last quarter (due to, say, unexpected climatic or economic conditions) will result in stock depletion, since contract renegotiation is gradual. We seek to capture this surprise term by comparing the previous quarter's demand with a four-quarter moving average.

The results sketched here summarize a larger effort (Hubbard and Weiner 1983). In the absence of any a priori wisdom concerning the correct structure for the regression, we use logarithms; the effect of changes can thus be readily compared across variables and the results interpreted as elasticities. The regression was estimated using two-stage least squares over the fourth quarter of 1974 (1974:4) to the second quarter of 1981 (1981:2) with all variables seasonally adjusted (the adjustment is based on quarterly data extending back to 1960) and in logarithmic form. The results were as follows (absolute values of t statistics in parentheses):

$$(I/S)_t - (I/S)_{t-1} = \underset{(2.0)}{.16} \left[\epsilon P_{t+1}^{US} - (1 + r_t) P_t^{US} \right]$$

$$- \underset{(4.0)}{.54} \left[(I/S)_{t-1} - (\overline{I/S})_{t-1} \right] - \underset{(5.5)}{1.44} \left[S_t - \overline{S}_t \right]$$

$$\overline{R}^2 = .54, \qquad D.W. = 1.24$$

where t indexes time, I is inventories, S is sales (both seasonally adjusted and in logarithmic form), P^{US} is the U.S. domestic price, r is the three-month Treasury bill rate divided by four, a bar over a variable indicates a four-quarter moving average, and ϵ denotes expectation. A description of expectation formation can be found in Appendix A.

The results imply that companies build stocks in terms of days of consumption (1) when they expect prices to increase by more than the cost of holding them, (2) when their inventory-to-sales ratios are below historical levels, and (3) when consumption is unexpectedly low. The coefficients, which are significant at the 95 percent level, can be interpreted as elasticities. A 1 percent increase in expected profit leads to a 0.16 percent increase in the inventory-to-sales ratio, or roughly 20 thousand barrels per day (inventory-to-sales ratio of 80 days $\times$ 0.16 percent $\times$ 15.5 mbd sales $\approx$ 2 mb/quarter $\approx$ 20 thousand barrels per day). It is too simplistic to characterize inventory accumulation as "procyclical" or "anticyclical" per se. Stock build or stock draw (or both) can accompany the various stages of an oil supply disruption.

POLICY ISSUES

The various issues surrounding stocks policy can be distilled into four questions: the size of the reserve, the fill schedule, the draw schedule, and the institutional framework (e.g., ownership, finance, decision-making). Clearly, the answer to any one of these questions is constrained by the answers given to the other three. However, devising a comprehensive cradle-to-grave program for the reserve is prohibitively complicated, the answer to each of the questions being far from trivial. It is also of limited value to policymakers, who seldom, if ever, have the luxury of creating the world anew.

In this chapter, we concentrate on drawdown decisions and ignore the questions of size, fill schedule, and institutional framework. Our reasons for doing so are two. First, the drawdown issue has received considerably less analysis than the others.[12] Second, while most economists tend to agree, or at least do not disagree strenuously, about the size of the reserve and its institutional setup, considerable controversy exists about its use as an instrument of policy.[13] It should be obvious that a reserve that is never intended to be used is no better than no reserve at all.

In analyzing drawdown alternatives, a firm grasp of the objectives of policy is paramount. The adverse economic effects (losses in GNP, increased unemployment, and inflation) of a supply shock such as a rapid oil price increase are well known; we take the goal of stockpile release as mitigation of this damage through the above-described exertion of downward pressure on oil prices. We regard debates over whether reserves should be used for strategic or tactical purposes as vacuous; we assume that in the event of war, the armed forces will have priority access to oil as well as other goods. We assume that government reserves are instruments of energy emergency policy, not a means for manipulating oil prices in the medium run. As David Stockman (1980) observed "That would be a little like a mouse taking on an elephant; our [proposed OECD reserves of] two billion barrels in salt domes is no match for their 500 billion barrels in the ground."

As noted above, an inherent problem in any contingency program is determining when the emergency has occurred. Below, we devote some attention to this critical question of timing, which lies at the heart of the subtrigger controversy.

Options

What alternatives—short of pulling the trigger—are available to OECD members when confronted with the specter of a disruption? First, they can do nothing. Laissez faire has been a popular option in the past and can serve as a control for "do something" alternatives.

In the "do something" sphere, recent stress appears to be on informality, flexibility, and consultations among governments and between governments and oil companies (see Becker 1982; Keohane 1982; and Chapter 4). Such measures, while admirable as modi operandi, are of little value in the absence of ideas to talk about. Otherwise, they will very likely wind up serving as a fig leaf for a "do nothing" policy.

Second, governments can exhort companies to refrain from making spot market purchases, or at least from doing so at "abnormal" prices. We regard spot prices as being determined by supply and demand. Thus, insofar as spot market abstinence follows from stock draw or demand restraint, price pressure will be eased. Otherwise, such exhortations are foolish. The spot price represents the thermometer, not the illness, and disabling the former can do little for the latter.

Third, due to rigidities in the oil market, a supply shock is likely to affect countries unequally. Informal reallocation can correct these imbalances and, hence, limit recourse to the spot market. This option is thus a version of the last; reallocation can be effective if accompanied by demand restraint or stock draw.

Fourth, as discussed above, demand restraint is effective, but would be terribly difficult to implement. The fifth alternative is that of "flexible stocks"—our focus for analysis in this chapter. Several approaches are possible. One plan of attack is to calculate the optimal draw-cum-fill schedule for the SPR under the various assumptions about the stock behavior of other OECD members. Such a procedure entails a dynamic programming solution and requires that the oil market be characterized by a small number of discrete states, with known probabilities of going from one state to another. Teisberg (1981) took this approach; Hogan (1982) extended it to incorporate international interaction. Neither considered private inventory behavior. Wright and Williams (1982) incorporated private stocks, but their assumptions result in inventory draw in a disruption; the behavior examined here is excluded. Another possibility is to assume away most of the uncertainties associated with the disruption and then optimize by linear programming methods. Such an approach was taken by Kuenne, Blankenship, and McCoy (1979).

Despite its attractive features, we do not employ an optimizing model in this chapter. The data and simplifying assumptions required to do so limit its value in short-term policy applications, Instead, we adopt the less intellectually satisfying but simpler approach of comparing a few proposed strategies designed to be illustrative of a broad range of alternatives.

We examine five stock policies:

1. Laissez faire.   In this "base case," countries rely on the market to allocate supplies. The IEA mechanism is not triggered, even in a severe disruption.
2. The United States relies on the market, other countries release five days of stocks in the notation used above, $(I/S)_{t+1} - (I/S)_t = -5$) held compulsorily. A plan of this type has been advanced by the Commission of the European Communities (1981).
3. As in 2., but other countries increase stockpiles by five days (i.e., $(I/S)_{t+1} - (I/S)_t = 5$).
4. Unilateral U.S. SPR drawdown of five days of consumption (other countries rely on the market).
5. All countries participate in an agreement to release from stocks the equivalent of five days of consumption.

The players involved are three: the U.S. SPR authorities, the U.S. oil industry (whose behavior is described above), and the foreign authorities. Diverse forms of regulation make separation of public and private stock behavior abroad a formidable task and prevent our treating foreign stock behavior endogenously. This is equivalent to assuming either that foreign governments are able to dictate stock behavior to their domestic oil companies or that foreign stock changes reflect decisions made in both the public and private sectors. Neither is particularly palatable, but modeling foreign stock behavior must await future work.

Economic Analysis of Stock Policy Alternatives

The economic effects to be considered are two. First, we seek to quantify the above-described effects of stock adjustments on the world oil market. Second, we examine the resulting macroeconomic impact, which is presumably the raison d'être of stocks policy. The effects of unilateral American action (e.g., the infamous 1979 middle distillate entitlement) on other buyers in the world oil market are well appreciated. In this chapter, we turn the question around by investigating the impacts of OECD policies on the U.S. economy, utilizing a short-run macroeconomic model linked to a model of the world oil market. Unlike most large macroeconomic models (wherein the oil price is a datum), the linkage here runs in both directions. The world oil market model and the macroeconomic model are solved simultaneously; income effects of oil price changes feed back into oil demand, which affects the price of oil, and so forth. A brief description of the macroeconomic model can be found in Appendix A; consult Hubbard and Fry (1982) for more detail and documentation.

How can policies designed to influence oil prices affect the economy? Through which channels do oil shocks have economic impacts? First, the increased oil price diverts spending from home-produced goods to imports, increasing the transfer of resources to oil-producing countries and reducing aggregate demand for U.S. output (net exports fall since oil exporters don't spend all their increased wealth immediately). Second, the rise in the relative price of oil, an important input, reduces the profit-maximizing level of output for firms that use oil, necessitating a fall in real GNP from the supply side. This reduction in output reduces the demand—at least in the short run—for other inputs, such as labor, thus lowering the real wage at which the supply of and demand for labor would be equalized.

These direct aggregate demand and supply effects are magnified because the economic system is not perfectly flexible. Because of rigidities in the economy, particularly sticky real wages, unemployment results, and the economy will fail to attain its (already diminished) consumption and production possibilities. The failure of wages and prices to adjust downward aggravates the rise in the price level caused by an oil price increase. The ultimate consequences for inflation and real income will depend on the size and timing of the disruption (and oil price increases), on the effect of the consequent price level increase on wage settlements and on the fiscal, monetary, and regulatory responses of the government.

In principle, two types of policy responses could be advanced to reduce the economic costs of a large oil price increase attendant to a supply disruption: income stabilization policies and oil market policies. The former include such devices as temporary tax rebate schemes, accommodating monetary policy, or investment incentives — tools for mitigating the short-run drain on aggregate demand for a given oil price increase. Oil market policies encompass oil taxes and tariffs, price controls, and stockpile drawdown programs — interventions designed specifically to alter oil prices directly. Clearly, the two categories are not mutually exclusive.

There are two basic difficulties with relying solely on stabilization policies to address the problems of oil shocks. First, precisely to the extent to which they are successful (in maintaining real income), oil demand remains high, keeping upward pressure on prices.[14] Second, the imprecision with which stabilization policies achieve their goals is well known. The lags inherent in implementing policy, and phasing it out — not to mention correctly diagnosing the problem — may well nullify the intended result. Temporary tax changes may be ineffective. An accommodating monetary policy runs the risk of raising inflationary expectations.

Analyses of oil market policies have produced similarly ambiguous conclusions. Oil taxes and tariffs put downward pressure on world oil prices, but in the process, they may exacerbate the price shock at home and generate a substantial fiscal drag problem. Oil price controls reduce the domestic price, but they may reduce incentives for domestic production and increase the demand for imported oil, thereby raising world oil prices. In contrast, stockpile releases can serve to reduce world oil prices without the unpleasant economic side effects.

SIMULATION RESULTS

Our simulations cover the five-year period 1982 through 1986; results for 1986 are omitted for brevity. We first present a control scenario, wherein no further disruption takes place. Combined Iranian and Iraqi production recovers throughout the interval. U.S. oil production is projected to drop slowly throughout the interval, although it is more than offset by increases outside OPEC. We expect a relatively loose market (i.e., low output-to-capacity ratio) in the absence of further deliberate restrictive actions by OPEC, with oil prices first falling and then rising as the OECD economies recover from the recession. In the control scenario, the United States fills the SPR at 200 thousand barrels per day.

The Disruption

We simulate a disruption of moderate size. Available capacity is reduced by 7 mbd for one year, starting in 1983:1. Much of the loss is made up (albeit at higher prices) because of the substantial excess capacity at that time. (In the absence of a disruption, capacity would be utilized at about 75 percent.) With no policy intervention—we actually assume that the U.S. government intervenes to the extent of halting SPR fill during 1983—(endogenous) OPEC production falls relative to the base case roughly by 100 thousand barrels per day in 1983:1, 400 thousand barrels per day in 1983:2, 800 thousand barrels per day in 1983:3, 1.4 mbd in 1983:4, 2.2 mbd in 1984:1, 2.4 mbd in 1984:2, 2.9 mbd in 1984:3 and 3.6 mbd in 1984:4.

Table 11-1 and Figure 11-2 present quarterly spot and U.S. refiners' acquisition cost data and changes in private inventories. Although capacity is restored by 1984:1, the spot price does not descend to its predisruption level until 1985:1. U.S. private inventories are built relative to the control, rapidly at first (roughly 90 million barrels in 1983), more slowly thereafter. Starting in 1984:4, private inventories are decumulated relative to the control.

Policy

Borrowing from the European Communities Commission proposal, we examine a policy of releasing from inventory five days worth of consumption. U.S. and rest-of-OECD consumption are taken as roughly 16 mbd and 20 mbd, respectively, which translates into releases of 80 and 100 million barrels.[15] We assume these releases are initiated in the second quarter of the disruption (1983:2) and are repeated in the third quarter, amounting to a 900 thousand barrels per day SPR draw and a 1.1 mbd draw of foreign stocks. The 160 million barrels released from the SPR represent about half of its current contents. Obviously, other configurations are possible. Table 11-2 presents the effects on spot prices. Table 11-3 presents the effects on U.S. real GNP.

Table 11-1. Comparison of Control and Base Case Disruption.

Quarter	Spot Price (U.S. $/barrel) C^a	Spot Price (U.S. $/barrel) D^b	Refiners' Acquisition Cost (U.S. $/barrel) C	Refiners' Acquisition Cost (U.S. $/barrel) D	U.S. Private Stock Change (thousands of barrels per day) C	U.S. Private Stock Change (thousands of barrels per day) D
1982: 1	33.15		35.20		-840	
2	33.05		35.70		620	
3	32.30		35.85		550	
4	32.25		35.90		-410	
1983: 1	31.05	43.90	35.65	39.55	-1050	-750
2	29.70	51.85	35.00	44.60	370	580
3	27.80	55.70	34.00	49.40	500	730
4	29.80	66.50	33.90	56.10	-290	- 40
1984: 1	28.20	57.95	33.40	58.25	-1110	-1110
2	27.45	50.80	32.80	57.60	230	350
3	27.30	44.65	32.30	55.25	450	520
4	27.80	39.20	32.15	51.85	-330	-410
1985: 1	28.20	34.15	32.15	47.90	-1030	-1210
2	29.00	30.30	32.40	43.90	250	150
3	30.10	27.35	32.90	40.20	480	370
4	31.90	25.25	33.85	36.90	-240	-420

a. C = control.
b. D = disruption.

The tables reveal that given a U.S. laissez faire policy (no SPR draw or fill) during the disruption, it makes a noticeable, although not enormous, difference whether the rest of the OECD follows a cooperative (draw) or noncooperative (build) path. Even under the moderate assumptions employed—the 200 million barrel stock increase or decrease is roughly equal to the rest-of-OECD build in 1979 (U.S. Department of Energy 1983:100)—the difference in the spot price is substantial: about $7 per barrel in the first "policy" quarter (1983: 2), $11.75 in the second (recall that the policy lasts only two quarters), $8.60 in the third, and about $7 in the fourth. By 1985 the effect on the spot price is insignificant, but U.S. domestic prices (not shown) remain higher (by about $4 per barrel in 1984: 4).

Figure 11–2. Oil Prices: Control versus Disruption.

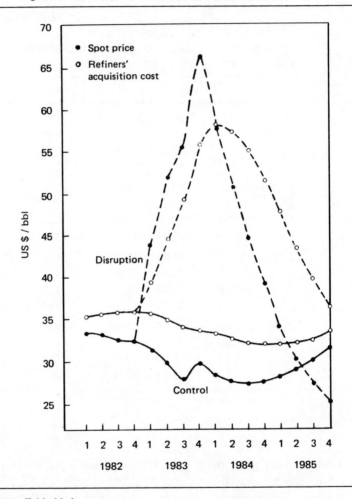

The loss in U.S. GNP is on the order of 30 to 40 percent greater in the noncooperative case in the first two quarters, 15 to 20 percent greater in the next two quarters.

The effects of using the SPR are two. First, the decreased demand for OPEC output exerts downward pressure on spot prices. It can be seen from Table 11-3 that this beneficial effect is only about three-fourths as large as when foreign stocks are decumulated, due to the

Table 11-2. Effects of Stock Policies on Spot Prices (*in U.S. $/barrel*).

Quarter	Spot Price Base Case Disruption	Difference in Spot Price Relative to Base Case			
		U.S.-Market Rest-Build	U.S.-Market Rest-Draw	U.S.-Draw Rest-Market	All Draw
1983: 1	43.90	0	0	0	0
2	51.85	3.70	-3.25	-2.55	-5.45
3	55.70	6.25	-5.50	-4.25	-9.20
4	66.50	4.60	-4.00	-2.95	-6.55
1984: 1	57.95	3.80	-3.30	-2.40	-5.30
2	50.80	3.00	-2.60	-1.85	-4.15
3	44.65	2.30	-1.95	-1.40	-3.15
4	39.20	1.65	-1.35	- .95	-2.15
1985: 1	34.15	1.05	- .85	- .70	-1.25
2	30.30	.50	- .35	- .20	- .50
3	27.35	.05	.05	.10	.20
4	25.25	-.40	.45	.35	.85

Table 11-3. Losses in U.S. Real GNP (in billions of 1982: 1 $/year relative to control) under Various Stock Policies.[a]

Quarter	Base Case Disruption	U.S.-Market Rest-Build	U.S.-Market Rest-Draw	U.S.-Draw Rest-Market	All Draw
1983: 1	1.6				
2	5.6	6.6	4.8	-5.0	-5.6
3	11.8	13.6	10.2	-1.4	-2.8
4	20.0	21.8	18.2	17.0	15.4
1984: 1	24.2	26.0	22.4	22.4	20.6
2	30.4	32.2	28.4	28.0	26.2
3	34.6	36.4	32.8	32.2	30.4
4	36.4	38.0	34.8	34.2	32.8
1985: 1	36.6	37.6	35.2	35.2	34.0
2	39.0	40.8	37.2	37.8	35.8
3	40.8	43.0	38.6	39.2	37.2
4	38.8	41.2	36.6	37.2	35.0

a. Negative numbers indicate GNP is higher than in the control.

smaller size of the SPR draw (900 thousand barrels per day vs. 1.1 mbd). Second, SPR draw substitutes for imported oil almost entirely; U.S. oil imports fall by 900 thousand barrels per day in 1983: 2 and 850 thousand barrels per day in 1983: 3. This import reduction improves the U.S. trade balance and hence the U.S. GNP. Given that market policies are followed by the other OECD members, SPR drawdown in 1983: 2 and 1983: 3 reduces the loss in real GNP by $7.2 billion (all GNP figures are in 1982: 1 dollars) over the year starting in 1983: 2. Since the disruption costs the economy $21.5 billion during this period, use of the SPR recoups about one-third of the loss. A rough estimate of the value of SPR oil can be obtained by dividing the $7.2 billion by the 160 million barrels released, yielding $45 per barrel. This figure measures economic benefits over and above the revenues accrued from SPR sales. In an intertemporal optimization calculation, these revenues must be compared with the costs (which we also ignore) of buying and storing the oil for the reserve.

Finally, we compare the cooperative (all draw) case with laissez faire: The 2 mbd decumulation shaves about $5.50 off the spot price in 1983: 2 and just over $9 in 1983: 3. In the following two quarters, spot prices are $6.55 and $5.30 lower, respectively. U.S. GNP is increased over the year starting in 1983: 2 by $8.2 billion—about 20 percent more than in the unilateral drawdown case.

CONCLUSION

Flexible stock policies, including "don't build," can play a significant role in moderating spot price fluctuations and reducing disruption-induced losses in real income, even in a moderate (subtrigger) disruption, wherein the U.S. refiners' acquisition cost rises by about 60 percent (50% in real terms). Inflation and unemployment (not shown) are also reduced. Cooperation provides noticeable tangible benefits in addition to the widely claimed spiritual advantages. Benefits are understated insofar as GNP is an imperfect measure of economic welfare because it ignores wealth transfers abroad. The "terms-of-trade adjusted" figure would be larger. Still, we should note that while the gains from stockpile coordination are significant, they are hardly overwhelming. Whereas macroeconomic policy responses are partly responsible for damage associated with past oil shocks, we have held them constant throughout. In this respect, international coordination of stabilization policies can play a role.

We find the claim that government stock drawdown is impotent due to countervailing actions taken in the private sector to be unfounded. Government releases dampen spot price increases, serving to reduce private inventory accumulation, not increase it. Unless the structural forces behind company decisions are different from in the past, private stockpiling will tend to exacerbate the disruption effects in the early stages and mitigate them thereafter.

Our simulations are designed to be illustrative, and they clearly depend on our policy assumptions. We are also constrained by the well-known limitations of econometric models, but we nevertheless believe that until policy is truly conducted on an optimizing basis, this type of analysis will continue to be a useful input to policy discussions.

NOTES TO CHAPTER 11

1. The Agreement on an International Energy Program, which set up the IEA, was signed by sixteen of the twenty-four OECD countries in November 1974. At this writing, twenty-one countries are members, including all of the major countries but France.

2. For detailed description and analysis, see Keohane (1982), Krapels (1980), U.S. General Accounting Office (1981), and Weiner (1981). For an official view, see Lantzke (1978) and Chapter 4.

3. The agreement can also be triggered when one member country loses 7 percent of its base period consumption but the IEA as a whole does not. Given the rigidities present in the oil market, this "specific trigger" situation is far more likely to occur than the "general trigger" described above, wherein countries are assumed to be affected equally. Given, however, that more oil is always available to those willing to pay, it is difficult to assign a meaningful interpretation to this case, and it will not be discussed further.

4. A 12 percent shortfall requires members to cut consumption by 10 percent. In either case, stockpile draw is pro rata on the basis of imports; thus, more self-sufficient countries have smaller drawdown obligations.

5. This will be strictly true only if the elasticities of IEA and free world demand are equal.

6. For a discussion of the subtrigger crisis and speculation on why demand restraint was inadequate in 1979, see Lantzke (1982).

7. For details, see U.S. Department of Energy (1981b: 29), wherein the door is left open: "The Administration continues to support close consultation with IEA members and would not rule out future actions in the event of [a small] interruption, depending on circumstances."

8. U.S. Department of Energy (1981b) sets out the official emergency policy of primary reliance on the market and substitution of SPR releases for government intervention to allocate supplies.

9. Authority to draw down the SPR stems from the Energy Policy and Conservation Act of 1975. The president must find that "a severe energy supply interruption or obligations of the United States under the International Energy Program" necessitates SPR release, where "severe" is defined as "of significant scope and duration, and of an emergency nature" and causing "major adverse impact on national safety or the national economy." This information is taken from Allen (1982). The definitions of "significant scope and duration" and "major adverse impact" are left open.

10. We use data on Mideast Light, since it makes up the largest fraction of crude oil traded internationally. Our results will thus be biased to the extent that the behavior of spot-contract price differentials varies systematically across grades of crude oil. For further discussion, see Verleger (1982).

11. This description of the short run is consistent with several views of medium- to long-term behavior, and we are thus agnostic on the question of OPEC internal structure, whether it be a cartel, a dominant-country oligopoly, or something else.

12. The size question has been addressed by Balas (1980), Teisberg (1981), Hogan (1982), and Chao and Manne (1982). Institutional issues are treated by Krapels (1980, 1982), and Plummer (1982). Kuenne, Blankenship, and McCoy (1979) examine drawdown decisions. Chapter 9 reviews the present status of U.S. stockpile policy.

13. Neither do we address the questions of how the SPR oil should be sold or to whom it should be sold. For a discussion of these issues, see Allen (1982). For a specific proposal to sell SPR oil through a futures market, see Chapter 10.

14. As mentioned above, there are certain "necessary" costs of an oil supply shock. If stabilization policies attempt to completely insulate the economy, inflation will be higher than it need be in the short run, and the economy's long-run growth path may be adversely affected.

15. These consumption figures are for stock release illustration only. In the model, consumption is endogenous.

REFERENCES

Allen, Myron. 1982. "Strategic Petroleum Reserve Drawdown and Distribution Decisions." In *Symposium Papers from Energy Modeling IV: Planning for Energy Disruptions.* Chicago: Institute of Gas Technology.

Bales, Egon. 1980. "Choosing the Overall Size of the Strategic Petroleum Reserve." In *Energy Policy Modeling: U.S. and Canadian Experiences*, Volume I, edited by W. Ziemba, S. Schwartz, and E. Koenigsberg, pp. 144-158. The Hague: Nijhoff Publishing.

Becker, Elmar. 1982. "The Use of Oil Stockpiles: The View from Bonn." In *International Oil Supplies and Stockpiling: Conference Proceedings*, edited by E.N. Krapels, pp. 38-41. London: Economist Intelligence Unit.

Chao, Hung-Po, and Alan S. Manne. 1982. "An Integrated Analysis of U.S. Oil Stockpiling Policies." In *Energy Vulnerability*, edited by J.L. Plummer, pp. 59-82. Cambridge, Mass.: Ballinger Publishing Company.

Chao, Hung-Po, and Stephen C. Peck. 1980. "Coordination of OECD Oil Import Policies: a Gaming Approach." Electric Power Research Institute. Mimeo, September.

Commission of the European Communities. 1981. "Measures to Limit the Effects of a Limited Shortfall in Oil Supply." COM (81) 533. Mimeo, September.

Danielsen, Albert L., and Edward B. Selby. 1980. "World Oil Price Increases: Sources and Solutions." *Energy Journal* 1, no. 4 (October): 59-74.

Frankel, Paul H. 1982. "Motivations of Governments and Companies." In *International Oil Supplies and Stockpiling: Conference Proceedings*, edited by E.N. Krapels, pp. 3-6. London: Economist Intelligence Unit.

Hogan, William W. 1982. "Oil Stockpiling: Help Thy Neighbor." Harvard Energy Security Program Discussion Paper Series, H-82-02.

Hubbard, R. Glenn, and Robert C. Fry. 1982. "The Macroeconomic Impacts of Oil Supply Disruptions." Energy and Environmental Policy Center Discussion Paper Series, E-81-07, Kennedy School of Government, Harvard University.

Hubbard, R. Glenn, and Robert J. Weiner. 1983. "Oil Inventory Behavior: An Empirical Analysis of Public-Private Interaction." Harvard Energy Security Program Discussion Paper Series, H-83-02.

Keohane, Robert O. 1982. "International Agencies and the Art of the Possible: The Case of the IEA." *Journal of Policy Analysis and Management* 1, no. 4 (Summer): 469-81.

Krapels, Edward N. 1980. *Oil Crisis Management*. Baltimore: Johns Hopkins University Press.

Krapels, Edward N., ed. 1982. *International Oil Supplies and Stockpiling: Conference Proceedings*. London: Economist Intelligence Unit.

Kuenne, Robert E.; Jerry W. Blankenship; and Paul F. McCoy. 1979. "Optimal Drawdown Patterns for Strategic Petroleum Reserves." *Energy Economics* 1, no. 1 (January): 3-13.

Lantzke, Ulf. 1978. "The International Energy Agency." *European Yearbook* 26: 41-65.

_____. 1982. "The Use of Oil Stockpiles: The View from the International Energy Agency." In *International Oil Supplies and Stockpiling: Conference Proceedings*, edited by E.N. Krapels, pp. 46-52. London: Economist Intelligence Unit.

Manne, Alan S. 1982. "The Potential Gains from Joint Action by Oil Import-ing Nations." In *Energy Vulnerability*, edited by J.L. Plummer, pp. 233–55. Cambridge, Mass.: Ballinger Publishing Company.

Nordhaus, William D. 1980. "Oil and Economic Performance in Industrial Countries." *Brookings Papers on Economic Activity*, no. 2: 341–88.

Plummer, James L. 1981. "Methods for Measuring the Oil Import Reduction Premium and the Oil Stockpiling Premium." *Energy Journal* 2, no. 1 (Jan-uary): 1–18.

_____ . "U.S. Stockpiling Policy." In *Energy Vulnerability*, edited by J.L. Plum-mer, pp. 115–148. Cambridge, Mass: Ballinger Publishing Company.

Rowen, Henry S., and John P. Weyant. 1982. "Reducing the Economic Impacts of Oil Supply Interruptions: An International Perspective." *Energy Journal* 3, no. 1 (January): 1–34.

Stockman, David. 1980. "Needed: A Dual Resurrection?" *Journal of Energy and Development* 5, no. 2 (Spring): 171–81.

Teisberg, Thomas J. "A Dynamic Programming Model of the U.S. Strategic Petroleum Reserve." *Bell Journal of Economics* 12, no. 2 (Autumn): 526–46.

U.S. Department of Energy. 1981a. "Securing America's Energy Future: The National Energy Policy Plan." DOE/S–0008. Washington, D.C. U.S. Govern-ment Printing Office, July.

_____ . 1981b. "Domestic and International Energy Emergency Preparedness." DOE/EP–0027. Washington, D.C.: U.S. Government Printing Office, July.

_____ . 1983. *Monthly Energy Review.* DOE/EIA–0035 (83/04) (April): 100.

U.S. General Accounting Office. 1981. "Unresolved Issues Remain Concerning U.S. Participation in the International Energy Agency." ID–81–38. Washing-ton, D.C.: U.S. Government Printing Office, September.

Verleger, Philip K. 1982. *Oil Markets in Turmoil.* Cambridge, Mass.: Ballinger Publishing Company.

Weiner, Robert J. 1981. "The Oil Import Question in an International Con-text: Institutional and Economic Aspects of Consumer Cooperation." Energy and Environmental Policy Center Discussion Paper Series, E–81–06, Kennedy School of Government, Harvard University.

Wright, Brian D., and Jeffery C. Williams. 1982. "The Roles of Public and Private Storage in Managing Oil Import Disruptions." *Bell Journal of Eco-nomics* 13, no. 2 (Autumn): 341–53.

MODELING THE WORLD OIL MARKET

This appendix provides some of the technical background of the economic model described in Chapters 10 and 11, emphasizing its applications to analysis of international strategic stockpile coordination and private oil inventory behavior.

The price-reaction function in the text for the spot market price of oil can be expressed in equation form as:

$$P_t = \psi P_{t-1} + \beta f (S_t / S_t^*) , \qquad (A\text{-}1)$$

where P is the spot price, t indexes the time period, f represents the function discussed in the text ($f' > 0$), ψ and β are parameters to be estimated, and S and S^* refer to OPEC output and capacity output, respectively.

Capacity decisions are assumed to be determined by longer term considerations outside the scope of the model and are taken as exogenous. S is obtained from the condition that world supply of and demand for oil must be equal:

$$DUS + DF + USS + FS = S + S^D + S^{NO} , \qquad (A\text{-}2)$$

where the variables are as defined in the following table.

Free World Oil Demand and Supply

Demand =		Supply =	
DUS (U.S. oil consumption)		S	(production by nondisrupted OPEC Members)
+ *DF*	(oil consumption in the rest of the free world)	+ S^D	(production by disrupted members)
+ *SPR*	(U.S. oil demand for the Strategic Petroleum Reserve)	+ S^{NO}	(non-OPEC oil production)
+ *USS*	(U.S. private oil inventory demand)		
+ *FS*	(inventory demand in the rest of the free world)		

U.S. consumption is assumed to depend on the domestic refiners' acquisition cost (P^{US}), income (Y^{US}), and a vector of structural variables, including past prices (X^{US}). Foreign consumption is defined similarly:

$$DUS = b^{US}(P^{US}, Y^{US}(P^{US}), X^{US}) \qquad \text{(A-3a)}$$

$$DF = b^F(P^F, Y^F(P^F), X^F) \ . \qquad \text{(A-3b)}$$

The U.S. refiners' acquisition cost is taken to be an average of spot and contract prices (plus transport costs), and to adjust to the spot price. The domestic price abroad is defined similarly:

$$P_t^{US} = a_0 + a_1 P_{t-1}^{US} + a_2 (P_t - P_{t-1}^{US}) \qquad \text{(A-4a)}$$

$$P_t^F = b_0 + b_1 P_{t-1}^F + b_2 (P_t - P_{t-1}^F) \ . \qquad \text{(A-4b)}$$

In general, the a's and the b's will be different for institutional and tax reasons. For now, consider the other variables as exogenous.

To obtain OPEC output in terms of consumption, stock change, and production by other countries, we can rearrange (A-2):

$$S_t = (DUS_t + DF_t) + (SPR_t + USS_t + FS_t) - (S^D + S^{NO}) \ . \qquad \text{(A-5)}$$

Production of oil by nondisrupted OPEC members is just world consumption demand plus world inventory demand less output of other producers. Substituting (A-5) back into (A-1) yields:

$$P_t = \psi P_{t-1} + \beta f \; \frac{(DUS_t + DF_t) + (SPR_t + USS_t + DFS_t) - (S_t^D + S_t^{NO})}{S_t^*} \; .$$

$$(A-6)$$

This approach to short-run oil price determination emphasizes energy–economy interactions. By linking the oil market model to a domestic macroeconomic model, not only will changes in oil prices affect economic activity (aggregate demand, potential output, and the price level), but economic activity at home and abroad influences oil prices. The processes are simultaneous. In the formulation implied by (A-6), given OPEC capacity, changes in non-OPEC production and in world (private and public) inventory accumulation also determine world oil prices.

INTERNATIONAL STOCKPILE COORDINATION

In terms of (A-6), the objective of an SPR policy during a supply interruption should be clear: to reduce the demand for OPEC output, thus moderating increases in the output-to-capacity ratio and reducing pressure on the spot price. The domestic and international interaction effects described in the text can be described by differentiating (A-6) with respect to SPR changes. Under the assumption of a negligible income effect, we have that

$$\frac{dP_t}{dSPR_t} = \frac{\beta f'}{S_t^*} \frac{dS_t}{dSPR_t} = \frac{\beta f'}{S_t^*} \left[\left(\frac{\partial DUS}{\partial P^{US}} \right) \left(\frac{\partial P^{US}}{\partial P} \right) \left(\frac{\partial P}{\partial SPR} \right) \right. \quad (A-7)$$

$$\left. + \left(\frac{\partial DF}{\partial P^F} \right) \left(\frac{\partial P^F}{\partial P} \right) \left(\frac{\partial P}{\partial SPR} \right) + \left(1 + \frac{\partial FS}{\partial SPR} \right) \right]$$

Substituting from (A-4) into (A-7) and labeling the international interaction effect, $\partial FS / \partial SPR$, as γ yields:

$$\frac{\partial P_t}{\partial SPR_t} = \underbrace{\frac{\beta f'}{S_t^*}}_{\substack{\text{direct} \\ \text{effect}}} \underbrace{\left[1 - \frac{\beta f'}{S_t^*} \left(a_2 \frac{\partial DUS}{\partial P^{US}} + b_2 \frac{\partial DF}{\partial P^F} \right) \right]^{-1}}_{\text{feedback effect}} \underbrace{(1 + \gamma)}_{\substack{\text{international} \\ \text{interaction effect}}}$$

$$\underbrace{\phantom{\frac{\beta f'}{S_t^*} \left[1 - \frac{\beta f'}{S_t^*} \right]}}_{\text{domestic effect}}$$

$$(A-8)$$

SPECULATIVE PRIVATE INVENTORY BEHAVIOR

The analysis of the sale of *SPR* oil through a crude oil futures market relies heavily on a description of the dependence of private oil inventory speculation on anticipated future profits. How does private inventory behavior fit into the short-run modeling framework described above?

As the simplest possible case, let world oil demand be a multiple of U.S. oil demand. We take *FS*, S^D, and S^{NO} as exogenous. Oil consumption outside the United States (*DF*) is set at δDUS. U.S. *work-in-process inventory demand* is assumed to be a fixed fraction of consumption, say ωDUS. *Speculative* inventory demand in the United States is a function of the expected profit from holding stocks. That is, it depends on the expected future price one period from now, $_tP^e_{t+1}$, less the current price P_t, and interest and storage costs, rP_t and cP_t, respectively.

$$USS_t = \omega DUS + j(_tP^e_{t+1} - (1 + r + c) P_t), \quad j' > 0.^1 \qquad \text{(A-9)}$$

We can now rewrite (A-6) as:

$$P_t = \psi P_{t-1} + \beta f \left(\frac{(1 + \omega + \delta) DUS_t + SPR_t + j(_tP^e_{t+1} - (1 + r + c)P_t) + FS_t - (S^D_t + S^{NO}_t)}{S^*_t} \right).$$

$$\text{(A-10)}$$

Hence, movements in expected future prices affect prices today by influencing inventory demand. A direct sale of *SPR* oil in the spot market reduces oil prices by reducing current demand for OPEC oil. A sale of *SPR* futures (say, for delivery in the next period) lowers expected future prices, lowering inventory demand and the spot price today. Comparing the relative effectiveness of the two strategies requires an empirical analysis, that is, a quantification of the parameters of the model.

The difference between the two strategies can be illustrated in the following example. Using (A-1) and (A-4) and ignoring foreign stockpile changes and the income effects of stockpile changes on the margin, equation (A-10) can be totally differentiated to obtain:

$$dP_t \left[1 - \frac{f'}{S_t^*} \left(a_2 (1 + \omega + \delta) \frac{\partial DUS}{\partial P^{US}} - j' (1 + r + c) \right) \right] \qquad \text{(A-11)}$$

$$= \frac{f'}{S_t^*} \left(dSPR_t + j' d_t P_{t+1}^e \right) \qquad ,$$

or

$$dP_t = \sum_{i=0}^{\infty} (j')^i \prod_{j=0}^{i} (X_{t+j}/Z_{t+j}) \, dSPR_{t+i} \qquad , \qquad \text{(A-12)}$$

where

$$X_{t+i} = \frac{g'}{S_{t+i}^*} \qquad , \qquad \text{(A-13)}$$

the spot price response to a required change in OPEC output (the *direct* effect), and

$$Z_{t+i} = 1 - \frac{g'}{S_{t+i}^*} \left\{ a_2 (1 + \omega + \gamma) \frac{\partial DUS}{\partial P^{US}} - b' (1 + r + c) \right\} \quad , \quad \text{(A-14)}$$

the *feedback*, or general equilibrium, response of spot prices to induced changes in demand. The greater is the OPEC price response or the response of speculative inventory demand to expected profits, the larger is the effect of the *SPR* change on spot prices. The greater is the general equilibrium price response, the smaller will be the ultimate effect on spot prices of changes in *SPR* accumulation.

We can now examine the effectiveness of direct drawdown of the *SPR* for spot market sales and guaranteed drawdowns next period (as per a futures market). Abstracting from effects on the variance of the distribution of expected future prices, the change in the spot price for a given drawdown today from the *SPR* is:

$$\frac{dP_t}{d(-SPR_t)} = \frac{-X_t}{Z_t} < 0 \quad . \qquad \text{(A-15)}$$

Hence, a contemporaneous drawdown of *SPR* oil will unambiguously reduce the spot price. The magnitude of the reduction depends on the two general equilibrium responses discussed earlier.

The effect on today's spot price of a drawdown next period *announced* today is:

$$\frac{dP_t}{d(-SPR_{t+1})} \bigg|_t = \frac{b' X_t}{Z_t} \left(\frac{-X_{t+1}}{Z_{t+1}} \right) < 0 \quad . \qquad \text{(A-16)}$$

If S^* remains the same over the two periods (i.e., if the disruption gets "no worse" or "no better"), (A-16) could be rewritten as:

$$\frac{dP_t}{d(-SPR_{t+1})}\Big|_t = -b'\left(\frac{X_t}{Z_t}\right)^2 \quad . \tag{A-17}$$

MODELING ENERGY-ECONOMY INTERACTION

The simulation results presented in the text were drawn from the Hubbard–Fry model of the U.S. economy and the world oil market, which merges the basic oil market model described above with a core macroeconomic model, solving them simultaneously. A brief description of the model follows; see Hubbard and Fry (1982) for more detail.

The transmission of an oil shock is modeled through a set of structural linkages. A sudden increase in the price of oil lowers the economy's potential output, reducing aggregate supply. Aggregate demand effects come through several channels. Personal consumption spending depends on permanent income and consumer wealth; an oil shock would lower consumption both by reducing the value of the existing capital stock (wealth effect) and by reducing current output (income effect). Business fixed investment depends on expected output and on the cost of capital services, which includes the cost of borrowed and equity funds as well as considerations of depreciation, investment tax credits, and the corporate income tax. Oil price increases influence capital spending through their impact on those channels. Housing and inventory investment decisions are also modeled. Oil shocks also affect the economy through the current account, though this impact is much larger in the short run than in the long run in the model because of the difference in short-run and long-run price elasticities of the demand for oil, the propensity to import of oil-producing countries, and exchange rate movements. Shocks affect unemployment through their impact on real output in conjunction with their impact on real unit labor costs.

The model also emphasizes the determination of wages and prices as an important transmission mechanism. A common problem in many macroeconometric models is the simultaneity of the determination of wages and prices. Increases in unit labor costs are certainly a factor in inflation, but workers presumably consider inflation when making nominal wage demands. The growth of (nominal) wages in the model depends on inflationary expectations and on the unemployment rate. Labor compensation depends on wages, fringe benefits, and employers' contribution to payroll tax programs (like social security and unemployment insurance).

Inflationary expectations depend on lagged inflation and on money growth. Hence, while oil price shocks may ultimately affect wage demands through their inflationary impact, the stance of monetary policy (i.e., whether or not to accommodate the shock) is important for the path of nominal and real wages after the shock. The implicit price deflator for the GNP is determined from information about unit labor costs, the cost of capital services, and the aggregate price of energy (determined from the world oil market model and from assumptions about the prices of coal and natural gas).

The macroeconomic model also contains a model of the domestic money market, focusing on the supply of and demand for money. Short-term interest rates from that model in conjunction with a term structure equation (influenced by the financing of government debt) yield long-term interest rates (which influence business fixed investment) and mortgage rates (affecting housing demand). Central bank decisions on the growth of the monetary base also affect inflationary expectations, with resulting impacts on wage rate and exchange rate determination.

The government can also affect the outcomes of the variables in the model through changes in fiscal policy (taxes and spending). Changes in payroll taxes affect labor compensation and the price of output; changes in corporate income taxes, the investment tax credit, or allowable depreciation rates affect investment. In analyzing the impact of fiscal policy, the model focuses on (1) the timing of the revenue and expenditure changes, (2) the components of aggregate demand affected, and their feedbacks to the rest of the model, (3) the inflationary consequences of the changes, and (4) the way in which the change is financed.

NOTE TO APPENDIX A

1. Equation (A-9) is derived from optimizing behavior in Hubbard and Weiner (1983). In empirically testing our inventory model, we used the variables given in the text. That is, the change in the (seasonally adjusted) inventory-

to-sales ratio was regressed on the expected one-period profit from holding stocks, departures of the inventory-to-sales ratio from trend, and departures of sales from trend. See Hubbard and Weiner (1983) for details.

REFERENCES

Hubbard, R. Glenn, and Robert C. Fry. 1982. "The Macroeconomic Impacts of Oil Supply Disruptions." Discussion Paper E-81-07, Energy and Environmental Policy Center, Kennedy School of Government, Harvard University.

Hubbard, R. Glenn, and Robert J. Weiner. 1983. "Oil Inventory Behavior: An Empirical Analysis of Public-Private Interaction." Harvard Energy Security Program Discussion Paper Series, H-83-02.

INDEX

227

ABOUT THE EDITORS

Alvin L. Alm served as director of the Harvard Energy Security Program from 1981 to 1983, when this book was prepared. Mr. Alm was assistant secretary for policy and evaluation of the U.S. Department of Energy from 1977–1979. He has held positions at the Office of Management and Budget, the Council on Environmental Quality, and the Environmental Protection Agency.

Robert J. Weiner is a doctoral candidate at the Harvard Business School and Department of Economics and a member of the Energy Security Program. He has held research positions at the Electric Power Research Institute and Brookhaven National Laboratory and taught economics and operations research at Harvard University. His current research interests are the economics and regulation of energy and minerals markets and their interaction with the economy.

ABOUT THE CONTRIBUTORS

Daniel B. Badger, Jr. is an administrator in the Oil Industry Division of the International Energy Agency. He was formerly an analyst at the U.S. Department of Energy and served as a special assistant to the U.S. Nuclear Regulatory Commission.

Michael C. Barth is a principal of ICF Incorporated in Washington, D.C. He was formerly deputy assistant secretary of the U.S. Department of Health, Education, and Welfare. Dr. Barth has taught economics at the University of Wisconsin and the City College of New York.

Edwin Berk is a senior associate at ICF Incorporated, where he specializes in environmental analysis and energy policy. Dr. Berk was formerly on the staff of U.S. Senator John Heinz and served on the faculty of Yale University.

E. William Colglazier is a professor of physics and director of the Energy, Environment and Resources Center at the University of Tennessee. Dr. Colglazier was formerly a research fellow at the Center for Science and International Affairs at Harvard University and a member of the Energy Security Program.

Shantayanan Devarajan is assistant professor of public policy at the John F. Kennedy School of Government, Harvard University, and a member of the Energy Security Program. Dr. Devarajan has served as a consultant to the World Bank on projects in Malaysia, Cyprus, and Cameroon.

R. Glenn Hubbard is assistant professor of economics at Northwestern University and a member of the Harvard Energy Security Program. He is a participant and modeler in the Stanford Energy Modeling Forum study, "Energy Shocks, Inflation, and Economic Activity." Dr. Hubbard has taught economics at Harvard University.

Barbara Kates-Garnick is a doctoral candidate at the Fletcher School of Law and Diplomacy, Tufts University. She was formerly a member of the Harvard Energy Security Program and has served in local government and as a consultant to the U.S. Department of Energy.

Steven Kelman is associate professor of public policy at the John F. Kennedy School of Government, Harvard University, and a member of the Energy Security Program. Dr. Kelman is the author of *Regulating America, Regulating Sweden* (1981), and *What Price Incentives?: Economists and the Environment* (1982).

Edward N. Krapels is president of Energy Security Analysis, Inc. in Washington, D.C. and a doctoral candidate at the School for Advanced International Studies, Johns Hopkins University. He is the author of *Oil Crisis Management* (1980) and *Pricing Petroleum Products* (1982), and is a former member of the Harvard Energy Security Program.

Ulf Lantzke has been the executive director of the International Energy Agency since its creation in 1974.

Robert S. Pindyck is professor of applied economics at the Sloan School of Management, Massachusetts Institute of Technology. He is the author or coauthor of several books, including *The Structure of World Energy Demand* (1979). Dr. Pindyck has been a consultant to the Federal Energy Administration and the World Bank.

James L. Plummer is president of Q. E. D. Research, Inc. in Palo Alto, California. He was formerly the director of the Energy Analysis Department at the Electric Power Research Institute and director of Corporate Economics at Occidental Petroleum. Dr. Plummer is the editor of *Energy Vulnerability* (1982).

Julio J. Rotemberg is assistant professor of applied economics at the Sloan School of Management, Massachusetts Institute of Technology. Dr. Rotemberg's current research interests include energy economics, macroeconomics, and monetary theory.